T5-BCJ-683

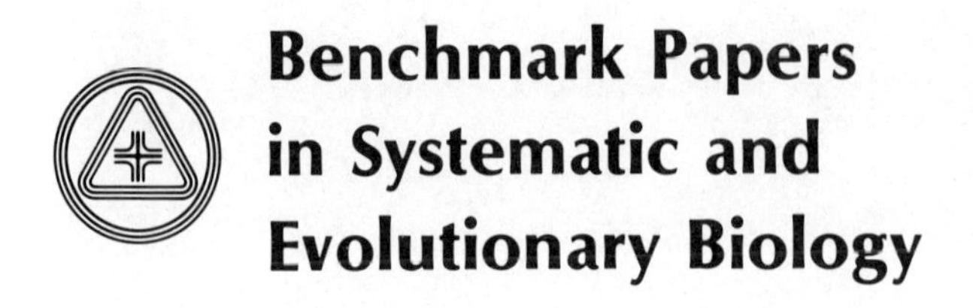

Benchmark Papers in Systematic and Evolutionary Biology

Series Editor: Carl Jay Bajema
Grand Valley State Colleges

Volume

1 MULTIVARIATE STATISTICAL METHODS: Among-Groups Covariation / *William R. Atchley and Edwin H. Bryant*
2 MULTIVARIATE STATISTICAL METHODS: Within-Groups Covariation / *Edwin H. Bryant and William R. Atchley*
3 CONCEPTS OF SPECIES / *C. N. Slobodchikoff*
4 ARTIFICIAL SELECTION AND THE DEVELOPMENT OF EVOLUTIONARY THEORY / *Carl Jay Bajema*

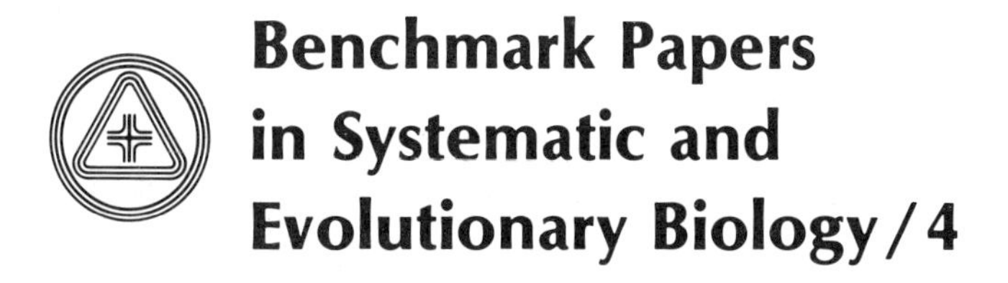

Benchmark Papers in Systematic and Evolutionary Biology / 4

A BENCHMARK® Books Series

ARTIFICIAL SELECTION AND THE DEVELOPMENT OF EVOLUTIONARY THEORY

Edited by

CARL JAY BAJEMA

Grand Valley State Colleges

Hutchinson Ross Publishing Company

Stroudsburg, Pennsylvania

Benchmark Papers in Systematic and Evolutionary Biology, Volume 4
Library of Congress Catalog Card Number: 80-10784
ISBN: 0-87933-369-3

84 83 82 1 2 3 4 5
Manufactured in the United States of America

LIBRARY OF CONGRESS CATALOGING IN PUBLICATION DATA
Main entry under title:
Artificial selection and the development of evolutionary theory.
(Benchmark papers in systematic and evolutionary biology; 4)
Includes index.
1. Evolution—Addresses, essays, lectures. 2. Breeding—Addresses, essays, lectures. I. Bajema, Carl Jay, 1937–
QH371.A77 575.01'6 80-10784
ISBN 0-87933-369-3

Distributed world wide by Academic Press,
a subsidiary of Harcourt Brace Jovanovich,
Publishers.

CONTENTS

SERIES EDITOR'S FOREWORD

The volumes of the *Benchmark Papers in Systematic and Evolutionary Biology* series reprint classic scientific papers on the evolution and systematics of organisms. These Benchmark Papers volumes do more than just provide scholars with facsimile reproductions or English translations of classic papers on a particular topic in a single volume. The interpretive commentaries and extensive bibliography prepared by each Benchmark Papers volume editor provide busy scholars with a review of the primary and secondary literature of the field from a historical perspective and a summary of the current state of the art.

This volume, the fourth in the series, is devoted to artificial selection and the role that it has played in the development of evolutionary theory. This volume is the first of six on selection theory being edited by C. Bajema. *Benchmark Papers in Systematic and Evolutionary Biology* volumes are also being planned for such topics as biogeography, comparative anatomy, comparative embryology, interspecific competition, Lamarckism, mimicry, mutationism, origin of the genetic code, orthogenesis, speciation, and theories of systematics.

CARL JAY BAJEMA

PREFACE

The theory of adaptive evolution by selection is the most important and the most widely misunderstood principle in biology. The six-volumes of Benchmark papers on selection theory that I am editing are designed to help students and scholars gain a better understanding of, first, the history of the controversy surrounding the relative importance of selection as an ecological process generating adaptive evolution, and, second, the current state of our scientific understanding of the role that selection plays in giving direction to evolution.

The theory that adaptive evolution occurs as the result of selection operating on heritable variations was proposed by Charles Robert Darwin and Alfred Russel Wallace in 1858. *On the Origin of Species by Means of Natural Selection, or the Preservation of Favoured Races in the Struggle for Life,* which Charles Darwin published the following year, triggered a scientific revolution with respect to our understanding of the origin and continuing evolution of species.

Charles Darwin was led to the idea that selection could bring about evolution by reading the literature on the breeding of domesticated plants and animals. It was only after Darwin had become convinced that selection was an ecological process that could cause "descent with modification" (evolution) in domesticated species that he discovered where the power of selection resided in nature. Darwin became convinced that selection was powerful enough to bring about evolution when he read *An Essay on the Principle of Population . . .* by Malthus (1826). Darwin concluded that the power of selection to bring about evolution in the rest of nature resided in competition, the "struggle for existence" produced by the overproduction of offspring.

The facsimile reproductions and English translations of the classic papers reprinted in this volume together with the commentary provide conclusive evidence that scientists who studied artificial selection have played a crucial role in the development of both the Darwinian and neo-Darwinian theories of adaptive evolution by natural, sexual, and artificial selection. It is impossible to publish all the important papers on artificial selection theory in one, 400-page volume. Classic papers on the role that artificial selection has played in the domestication of plants and animals will appear in a forthcoming volume in the *Benchmark Papers in Systematic and Evolutionary Biology* series. Many of the classic papers on selective plant and animal breeding that have been published since the rediscovery of Mendelian genetics (1900) will be reprinted in forthcoming *Benchmark Papers in Genetics* volumes on plant breeding and animal breeding.

This volume on artificial selection is the first of six Benchmark Papers volumes on selection theory. The next two will be devoted to natural selection and sexual selection. Domestication of plants and animals, human evolution by selection, and cultural evolution by selection will be the topics of the other three volumes on adaptive evolution by selection theory.

I wish to thank the many people who provided assistance while I selected papers for inclusion in this volume and wrote the editor's commentaries on the history of artificial selection theory: Viteszlav Orel, Mendelianum, Moravian Museum, Brno, Czechoslovakia; Will Provine, Cornell University; Roger Wood, University of Manchester; Peter Gautrey, University Library, Cambridge; Michael Ruse, University of Guelph; Whitfield Bell, American Philosophical Society, Philadelphia; A. S. Cain, University of Liverpool; Fred Churchill, Indiana University; and the librarians at Grand Valley State Colleges and elsewhere who located and obtained copies of numerous scholarly publications for me through interlibrary loans.

CARL JAY BAJEMA

CONTENTS BY AUTHOR

ARTIFICIAL SELECTION AND THE DEVELOPMENT OF EVOLUTIONARY THEORY

A long searching amongst agricultural and horticultural books and people makes me believe . . . that I see the way in which new varieties become exquisitely adapted to the external conditions of life and to other surrounding beings.

Charles Darwin, 1845
(Letter to L. Jenyns)

All my notions about how species change are derived from long-continued study of the works of (and converse with) agriculturalists and horticulturalists; and I believe I see my way pretty clearly on the means used by nature to change her species and adapt them to the wondrous and exquisitely beautiful contingencies to which every living being is exposed . . .

Charles Darwin, 1856
(Letter to Asa Gray)

You are right, that I came to the conclusion that selection was the principle of change from the study of domesticated production; and then, reading Malthus, I saw at once how to apply this principle.

Charles Darwin, 1859
(Letter to Alfred R. Wallace)

Part I

THE ART AND SCIENCE OF ARTIFICIAL SELECTION BEFORE CHARLES DARWIN

Editor's Comments on Papers 1 Through 8

What natural causes might account for the origin of species? Philosophers living in ancient Greece speculated about the answer to this question. Empedocles (495–435 B.C.) proposed the theory that the origin of all species occurred by spontaneous generation during a particular period of history. Individual organisms were spontaneously generated by the random combining of body organs that were already present in the environment. Necessity (selection) eliminated those individuals whose combination of body organs did not allow them to survive and reproduce (Osborn, 1894).

The Greek poet Theognis and the Greek philosopher Plato both expressed their concern over the hereditary quality of human beings by drawing an analogy between the careful selective breeding of domesticated species by humans and the way that mating and reproduction were occurring in human populations (Roper, 1913). Conscious human choice was involved in both of the two selective processes that Theognis and Plato compared. None of the extant writings of the Greeks contain statements about a possible analogy between the changes in domesticated species brought about by conscious human selection and processes operating in the rest of nature that could bring about selective change in wild species.

The Greek idea that came to dominate philosophical thought about the nature of species was Plato's belief in the existence of unchanging and eternal realities—ideal types (eidos)—that were supposed to be independent of the changing things of the world perceived by human senses. The widespread adoption of Plato's typological or essentialist view of nature by philosophers made it difficult, if not impossible, for them to seriously consider the possibility that the origin of species might be scientifically explained in terms of populations changing as the result of the accumulation of variations by selection (Mayr, 1959:2–4, 1972).

Francis Bacon (1561–1626), the intellectual father of modern scientific philosophy, considered the possibility that species could change as the result of the accumulation of variations. Bacon pointed out that humans take advantage of the chance variations of Nature and accumulate them by Art (artificial selection) in his 1620 book on the new machines of the mind, *Novum Organum* (Paper 1).

Bacon's thoughts on conducting scientific studies of change in Nature and in domesticated species were buried in his voluminous writings. They appear to have had little, if any, direct impact on the history of scientific theories with respect to evolution.

Bacon's ideas had to be rediscovered by historians of evolutionary thought (Osborn, 1894; Cook, 1932).

Francis Bacon's writings did influence Canon Robert Sharrock (1630–1684), who developed a scientific interest in the cultivation of plants and wrote *The History of the Propagation & Improvement of Vegetables by the Concurrence of Art and Nature* (Paper 2) in 1660. Sharrock was skeptical about the possibility of the transmutation of species even though some scholars believed that instantaneous evolutionary change could occur during reproduction. Sharrock tried to demonstrate how the belief in the instantaneous transmutation of species arose from insufficient scientific investigation. A greatly enlarged second edition of Sharrock's *History* was published in 1672.

The French mathematician and astronomer Pierre-Louis de Maupertuis (1698–1759) formulated the earliest comprehensive theory of transmutation based on mutation, selection, and geographic isolation. Using the analogy between the selective breeding of domesticated species and reproduction in the rest of nature, Maupertuis argued that if the ingenuity of man can produce species, why can't nature? Maupertuis first published his ideas in 1745 in *Venus Physique: Primiere Partie Contenant une Dissertation sur l'Origine des Hommes, et des Animaux; Second Partie Contenant une Dissertation sur l'Origine des Noirs* (Papers 3a and b), which he wrote in a popular style for gentlemen and ladies of the court in Berlin (Glass, 1959;62). Maupertuis published his most definitive statement concerning the role that he thought selection could play in the evolution of adaptations in his *Essai de Cosmologie* (1750):

> May we not say that, in the fortuitous combinations of the productions of Nature, since only those creatures *could* survive in whose organization a certain degree of adaptation was present, there is nothing extraordinary in the fact that such adaptation is actually found in all those species which now exist? Chance, one might say, turned out a vast number of individuals; a small proportion of these were organized in such a manner that the animals' organs could satisfy their needs. A much greater number showed neither adaptation nor order; these last have all perished. . . . Thus the species which we see today are but a small part of all those that a blind destiny has produced. (English translation in Glass, 1959:57–58)

Maupertuis' theory that an offspring can inherit from each parent more than one "germ" (particle or gene) for qualitative characteristics such as albinism and supernumerary (six) digits led him to conclude that in the absence of art or discipline the newly-produced variations would be overwhelmed by old variations in a few generations because germs "for the parts whose origin is ancestral are the more abundant in the seeds [semen]" (Maupertuis, 1745: 157, 1966:79).

Comte de Buffon, the great French naturalist who was a young contemporary of Maupertuis, described *Venus Physique* as a book that "although very short, assembles more philosophical ideas than there are alltold in several great volumes on generation" (Buffon, 1769, quoted in Glass, 1959). The evolutionary implications of Maupertuis' hypotheses were given wide publicity in 1754 by Diderot, the French encyclopedist.

Nonetheless, Marpertuis' hypotheses concerning the possible processes by which transmutation could occur had no lasting positive impact on Buffon (Wilkie, 1956) or any other member of the scientific community (Glass, 1959). Why did the fertile ideas of Maupertuis on evolution die out so completely in the minds of the scientific community that a mere century later Charles Darwin was completely unaware of his views? The highly accurate and comprehensive hypothesis of transmutation proposed by Maupertuis in 1745 was rediscovered by historians in the twentieth century. The French historian of science Emile Guyenot has succinctly summarized the intellectual achievement of Maupertuis:

> Geneticist from the very start, Maupertuis sketched, in all, an evolutionary conception that was based at once on heredity, on the fortuitous production of variations, on a prophetic vision of the phenomena of selection and of preadaptation. We feel him infinitely closer to us than the great theoreticians of Transformism. Already he was reasoning more on the basis of facts than of conceptions a priori. He was one of the founders of Evolution and appears to us, in addition, as the most remarkable precursor of contemporary Mutationism. (Guyenot, 1941: 393; quoted in Glass, 1959:82)

Numerous scientific investigators attempted but failed to ascertain either exactly what was inherited or the Mendelian patterns of inheritance during the two centuries preceding the publication of Charles Darwin's *On the Origin of Species* (1859). Botanists alone had conducted more than 9,000 plant-hybridization experiments that Gärtner reviewed in 1849. These scientific plant breeders provided evidence in support of the transmutation of species by demonstrating that interspecific hybrid plants could be produced by artificial pollination. They also helped demonstrate the importance of both parents in the inheritance of characteristics by offspring. While they developed certain practical techniques such as artificial pollination, these and other scientific investigators were unable to provide practical plant and animal breeders with a particulate theory of heredity that could be used in breeding improved varieties of domesticated plants and animals.

A major revolution in the art of domesticated plant and animal breeding began during the eighteenth century when a few practical men started selecting breeding stock strictly on the basis of economic characteristics. Some livestock breeders, for instance, began contending that the "fancy points" (color and pattern of coat, shape and size of horns, and so on) that distinguished local breeds were of little or no economic importance.

Robert Bakewell (1725–1795), an English tenant farmer, was the most famous of the eighteenth-century livestock breeders. He produced animals by selective breeding that fattened for market earlier and that had a higher proportion of saleable flesh. Bakewell's success was due to the fact that he chose the economic characteristics like muscle or bone that he wanted to change and he bred those animals that possessed the desired features employing inbreeding whenever possible (Wood, 1973:233). Bakewell considered his hands more useful than his eyes for judging stock. George Culley, a champion of Bakewell's methods of selective breeding, contended that if people who stressed the importance of noneconomic characteristics that distinguished local breeds could: "be prevailed upon to make an experiment, they would most probably find, that excellence does not depend on the *situation* or *size* of *horns*, or on the colour of *faces* and *legs* but on *more essential properties* . . . " (Culley, 1807:vi).

The first published account of the selective breeding methods and successes of Robert Bakewell was written by Arthur Young in 1771 (Paper 4). Robert Bakewell's successful cattle-breeding methods were summarized by such English agricultural writers as William Marshall (1790) and William Youatt who wrote:

> Improvement had hitherto been attempted to be produced by selecting females from the native stock of the country, and crossing them with males of an alien breed. Mr. Bakewell's good sense led him to imagine that the object might be better accomplished by uniting the superior branches of the same breed, than by any mixture of foreign ones. . . . As his stock increased, he was enabled to avoid the injurious and enervating consequences of breeding *too closely* "in and in." The breed was the same but he could interpose a remove or two, between members of the same family. He could improve all the excellencies of the breed without the danger of deterioration; and the rapidity of the improvement which he effected was only equalled by its extent. (Youatt, 1834:191–192; see also Paper 10)

One of Bakewell's major contributions to selective breeding

practices was his systematic use of progeny testing (Wood, 1973). He achieved this by frequently hiring out his bulls and rams rather than selling them. This enabled him to mate his prize males with as many females as possible. Bakewell was then able to judge the progeny of his prize bulls and rams with different mates and in different environments and to buy back the progeny he wanted for his own herds. Bakewell was not the first breeder to hire out his prize males, but he was the first to do so systematically and to popularize it against considerable opposition (Wood, 1973). The best scholarly review of Bakewell's selective breeding techniques has been written by Wood (1973).

Increased agricultural yields were also reported by eighteenth-century plant breeders who selected their breeding stock solely on the basis of utilitarian criteria. Joseph Cooper, (1759–1840), an American, and his father were able to dramatically increase yields from crop and garden plants by selecting plants for breeding stock strictly on the basis of economic characteristics. Joseph Priestley summarized the selection experiments of the Coopers in a letter that was published in 1798 (Paper 5).

Erasmus Darwin, Charles Darwin's grandfather, published numerous quotations from Priestley's letter on the Coopers' plant-selection experiments in his *Phytologia: Or the Philosophy of Agriculture and Gardening* (Darwin, 1800:409–411). Joseph Cooper also wrote a letter describing his selection techniques in 1799 that was reprinted several times in America and Europe (Cooper, 1808). The Moravian breeders whose experiments and writings on plant and animal hybridization influenced Gregor Mendel were aware of Cooper's experiments (V. Orel, 1978: personal communication).

In 1799 the Englishman Thomas Andrew Knight (1759–1838) described his endeavors to produce improved varieties of plants (Paper 7). Knight conducted breeding experiments with the garden pea in an attempt to gain knowledge about heredity that could be used to produce improved varieties by selection. Knight was also hoping that his breeding experiments would provide evidence for superfoetation, the idea that an organism can have two male parents in addition to the female parent. Evidence demonstrating superfoetation would have provided support for the then widely-held belief that the characteristics of males often predominate in offspring as the result of excess pollen or semen at fertilization.

Knight's 1799 paper describing his garden pea breeding experiments was translated into German and published in *Oekonomische Hefte* during 1800. A detailed account of Knight's numerous breeding experiments can be found in *Plant Hybridization*

Before Mendel (Roberts, 1929). The influence that Knight's ideas had on plant breeding in Moravia has been reviewed by Orel (1978). Numerous European plant and animal breeders were also employing selective breeding to bring about an improvement in the economic characteristics of domesticated species.

Robert Bakewell, Joseph Cooper, and Thomas Knight are but three of the many breeders that were able to bring about significant hereditary improvement in numerous economic characteristics of domesticated species by employing artificial selection in a more scientific way than ever before. The successes and selection methods of these and other breeders have been chronicled in numerous books, pamphlets, and articles published in England (Young, Paper 4, 1811; Culley, 1807; Marshall, 1790; Youatt, 1831, Paper 10, 1837, 1847; Le Couteur, 1836) and in Europe (Lecoq, 1845; Gärtner, 1849; Fraas, 1852). Numerous references to selection experiments carried out by plant and animal breeders before Charles Darwin's time can be found in a variety of scholarly histories (Roberts, 1929; Zirkle, 1935; Wallace and Brown, 1956; Trow-Smith, 1957, 1959; Stubbe, 1972; Bokonyi, 1974; Orel, 1977, 1978). The selective breeding of domesticated species during Greek and Roman times has been discussed by Cato and Varro (Harrison, 1917) and Roper (1913). The role that selective breeding has played in the evolution of domesticated species is the topic of a forthcoming Benchmark volume entitled the *Domestication of Plants and Animals by Selection*.

The idea that human populations could be improved by deliberate selective breeding was discussed by numerous writers who noted the analogy between the breeding of domesticated animals and reproduction in human beings before the publication of the Darwin-Wallace theory of evolution by selection in 1858 (Maupertuis, Papers 3A and B; Malthus, Paper 6; Prichard, 1836-1847; Lawrence, Paper 8; Walker, 1839; Noyes, 1849; see Roper 1913 for a review of the ideas of Plato and other ancient writers).

Robert Thomas Malthus (Paper 6) considered the possibility that mankind might be improved by artificial selection in his 1798 *An Essay on the Principle of Population as It Affects the Future Improvement of Society with Remarks on the Speculations of Mr. Godwin, M. Condorcet, and Other Writers.* Malthus contended, first, that the means by which populations must be kept down to the level of the means of subsistence forms "the strongest obstacle in the way to any great future improvement of society," and, second, that the amount of hereditary improvement that could be achieved in mankind by "attention to breed" was small in contrast to Condorcet's optimistic views concerning the organic perfectibility of man. It is interesting to note that while Condorcet (1795) contended that organic perfectibility of the intellectual and moral faculties could be brought about by the inheritance of qualities acquired by education, Malthus responded by discussing selective breeding of domesticated species and then arguing that selection could effect little change in humans because variation was too limited.

Lectures on Comparative Anatomy, Physiology, Zoology and the Natural History of Man delivered at the Royal College of Surgeons in 1816, 1817, and 1818 by William Lawrence (Paper 8) contains a passage illustrating how the analogy between the selective breeding of domesticated animals and human reproduction was used to make people more aware of what "selections and exclusions similar to those so successfully employed in rearing our more valuable animals" could accomplish if applied to human beings. The analogy that Lawrence drew between domesticated animals and man was probably due to the influence of J. Blumenbach who popularized this approach to physical anthropology (Wells, 1971:352). Lawrence's major source for his views on variation, heredity, and the role of sexual selection was the first edition of James Prichard's *Researches Into the Physical History of Man* (1813). Lawrence's *Lectures* influenced Edward Blyth (1835) and Alfred Russel Wallace (McKinney, 1969; Wells, 1971:357–358). Charles Darwin cited Lawrence on sexual selection in *The Descent of Man and Selection in Relation to Sex* (1871:II, 357). The evidence indicates that Charles Darwin probably did not read Lawrence's *Lectures* until 1847. Darwin read *Intermarriage* by Alexander Walker (1839) within a year after he had constructed his theory of adaptive evolution by selection. Walker's chapter on selection contained extensive quotations from both William Lawrence and James Prichard on the application of selective breeding to human reproduction.

The theory that populations of the human species have undergone adaptive radiation becoming races as the result of natural processes that act in a fashion analogous to artificial selection was championed in a paper read before the Royal Society in 1813 by William Charles Wells, an American-born physician practicing in England:

> Those who attend to the improvement of domestic animals, when they find individuals possessing, in a greater degree than common, the qualities they desire, couple a male and female of these together, then take the best of their offspring as a

> new stock, and in this way proceed, till they approach as near the point in view as the nature of things will permit. But, what is here done by art, seems to be done, with equal efficacy, though more slowly, by nature, in the formation of varieties of mankind, fitted for the country which they inhabit. (Wells, 1818:435)

Wells considered diseases and climate to be the most important factors involved in the origin of human races. Charles Darwin was not aware of William Wells's theory until after the first edition of the *Origin* was published.

1

Reprinted from pages 237–239 of *The Works of Francis Bacon,* vol. VIII, ed. J. Spedding, R. Ellis, and D. Heath, Hurd and Houghton, New York, 1864

NOVUM ORGANUM, BOOK II

Francis Bacon

[*Editor's Note:* In the original, material precedes and follows this except.]

XXIX.

Among Prerogative Instances I will put in the eighth place *Deviating Instances;* that is, errors, vagaries and prodigies of nature, wherein nature deviates and turns aside from her ordinary course. Errors of nature differ from Singular Instances in this, that the latter are prodigies of species, the former of individuals. Their use is pretty nearly the same; for they correct the erroneous impressions suggested to the understanding by ordinary phenomena, and reveal Common Forms. For in these also we are not to desist from inquiry, until the cause of the deviation is discovered. This cause however does not rise properly to any Form, but simply to the latent process that leads to the Form. For he that knows the ways of nature will more easily observe her deviations; and on the other hand he that knows her deviations will more accurately describe her ways.

They differ in this also from Singular Instances, that they give much more help to practice and the operative part. For to produce new species would be very difficult; but to vary known species, and thereby produce many rare and unusual results is less difficult. Now it is an easy passage from miracles of nature to miracles of art. For if nature be once detected in her deviation, and the reason thereof made evident, there will be little difficulty in leading her back by art to the point whither she strayed by accident; and that not only in one case, but also in others; for errors on one side point out and open the way to errors and deflexions on all sides. Under this head there is no need of examples; they are so plentiful. For we have to make a collection or particular natural history of all prodigies and monstrous births of nature; of everything in short that is in nature new, rare, and unusual. This must be done however with the strictest scrutiny, that fidelity may be ensured. Now those things are to be chiefly suspected, which depend in any way on religion; as the prodigies of Livy; and those not less which are found in writers on natural magic or alchemy, and men of that sort; who are a kind of suitors and lovers of fables. But whatever is admitted must be drawn from grave and credible history and trustworthy reports.

XXX.

Among Prerogative Instances I will put in the ninth place *Bordering Instances;* which I also call *Participles.* They are those which exhibit species of bodies that seem to be composed of two species, or to be rudiments between one species and another. These instances might with propriety be reckoned among Singular or Heteroclite Instances; for in the whole extent of nature they are of rare and extraordinary occurrence. But nevertheless for their worth's sake they should be ranked and treated separately; for they are of excellent use in indicating the composition and structure of things, and suggesting the causes of the number and quality of the ordinary species in the universe, and carrying on the understanding from that which is to that which may be.

Examples of these are, moss, which holds a place between putrescence and a plant; some comets, between stars and fiery meteors; flying fish, between birds and fish; bats, between birds and quadrupeds: also the ape, between man and beast, —

"Simia quam similis turpissima bestia nobis;"

likewise the biformed births of animals, mixed of different species, and the like.

2

Reprinted from pages 1–4 and 28–32 of *The History of Propagation & Improvement of Vegetables by the Concurrence of Art and Nature*, T. Robinson, Oxford, 1660

THE

HISTORY

OF

Artificial propagation of Plants.

R. Sharrock

CAP. I.

Of Propagation by Seed.

Num. 1. *Of Propagation of Vegetables in general, with a Preface to the Discourse.*

THe Illuſtrious and Renowned Lord *Bacon*, in his Diſcourſe concerning the advancement of Learning, reckons it among the Deficients of Natural Hiſtory, *That the Co-operation of Man, with Nature in particulars, hath not been obſerved; and that in thoſe Collections which are made of Agriculture, and other manual Arts, there is commonly a neglect and rejection of Experiments, familiar and vulgar, which yet to the interpretation of Nature*, and which I ſhall adde, general profit, *do as much, if not more conduce, then Experiments of a higher quality.* The ſame noble Perſon, in his

B par-

partition of Philosophy, complains of the want of an Inventary of what in any subjects by Nature and Art is certainly, and may be undoubtedly wrought. *I* believe his Lordship hath had many of his minde in former, has now, and is likely to have in future ages; for amongst those few Writings extant on these Subjects, some prove altogether useless, as being so full of their natural Magick and Romantick Stories, that we know no more what to credit in those Relations, in the Natural, then what in civil History we may believe of King *Arthur*; *Guy* of *Warwick* in ours; or of *Hector* and *Priam* in the *Trojan* Story: Others elevated in their Fancies, write in a Language of their own, addressing their Discourse to the Sons of Art, speaking rather to amuse, than instruct, and prove like blazing Stars, that distract many, and direct few.

Many of those who would write for Universal Instruction, either know the things that might make up the matter of their History, but want the skill to draw up such an Inventary, as his Lordship requires, as common Tradesmen and Artisans; or else indeed are learned enough to draw up the writing, but stand aloof from the knowledge of most of the particulars therein to be ingrost; which is the ordinary case of us, such of us as have pretensions to Scholarship.

I being necessitated by my obligations and respect to a Person truly Noble, to give some account of the particular effects of Man, co-operating with nature, in the matter of our English Vegetables, as they are improved by Husbandmen and Gardiners, desire to undertake no more, but to give a sincere endeavour, That the way of the Artist be set down, and the effect of Nature thereon; in the first of which, I intend

tend my directions so plain, as if appointed for the instruction of some Artists rude and untaught Apprentice: and the second's, if not so homely, yet as easie and evident, being a little disgusted with any thing intended for the use of Philosophy, when overgarnished with Rhetorical Tropes, which like Flowers stuck in a Window for whatsoever intended (either cheat or ornament) certainly create a darkness in the place. *Behemenical*, *Paracelsian*, and such Phrase as many Alchimists use, I must for the same reason avoid.

In the drawing up the Inventary, I will study that it may be true in all parts, and not to mingle, according to the example of *Pliny*, *Weeker*, *Porta*, and many more, both Latine and English Writers, any false relation, without its distinguishing Character; and if it be not perfect, it shall be for want of skill, or present remembrance of particulars.

The end of the Artist is to Propagate and Improve: To propagate, is to multiply the individuals of each kinde: And to improve, is to bring them, being propagated, to a more then ordinary excellency and goodness. The ways of increasing the particulars of each kinde, are, 1. By Seed, 2. By off-set, taken from a Mother-Plant. 3. By laying the Branch of a growing Plant down into the Earth. 4. By bearing up a Soil to it. 5. By Stems set without roots. And lastly, By the various ways of grafting and insitions.

Concerning all these, as likewise the preservation and melioration of things propagated, I shall endeavor to enumerate what Plants may be increased by each of these ways, and to shew how the operation in each may be performed, and what the product is

B 2 that

that by nature thence ordinarily ensues : Definitions are hopeless in this matter, useless too, and it might be harmful . If I should define Sowing, to be the casting of Seed into the Earth, in such maner, and at such time, when in the surface of the bed the earth would so ferment, as might be proper to the explication and further germination of the Seed and increase of the Plant, there might a world of controversies arise about the particulars therein contained ; and yet all that is there would be useless, till the particular Plants, and the maner of the operation, and time required to the sowing of their Seeds be first declared : I shall therefore wave all such endeavors, and hasten to what may rather prove for use than pomp.

[*Editor's Note:* Material has been omitted at this point.]

N. 6. *Some other relations of transmutation , and the possibility of a change of ones species into another examined.*

I have often heard persons affirme, that they have sowed Barley, or some other grain , and in the ground the seed has been so altered as to send forth Oates insteed of corn, according to its own species. I am as yet farre from giving any assent to this their History. The Reasons why I disbeleive them are, first, because the Relators affirme whole fields to be thus varied, and that to one species (viz) of Oates, which is different from Barley in the straw , eare and grain

it

it selfe. Whereas in the variation of seed, in those vegetables, in which the change is undoubted, the colour only or some other easily alterable accidents, such as the sensible qualities are generally found are transmuted, and this transmutation ends not at all in another divers kind; but in severall small diversities of the same kind; The storyes of Wheat turned to Mustard-seed were as likely to be true, and is a fit parallell to create a right beleife of the true cause of the mentioned effect. Secondly, I knew a Gentleman who plowed a piece of land in the spring, and then sowed it not, but after it was harrowed and prepared for seed left it to its own Genius and nature to produce what it was inclined to: The Ground was off its own Nature apt to bring forth wild-Oates amidst the Corn, now in defect of Corn there grew as many wild-Oates unmixt from any other weeds, as the land could carry. This was tryed in a great peice of land, and much proffit was made of the Oates, the Gentleman having cut them green for Fodder *Anno* 1657.

My judgment therefore is, That the fallacy which befell my above named Relators was, that they mistook the cause of the production of the Oates mentioned; for to me it is much more easie to conceive, that by some evill accident, as it often happens (the seedcorn being corrupted and perish't in the ground) the ground it's self from its own Seminary, sent out the supposititious Crop of Oates or Mustard, than that there should be a variety of so strange a Nature, and declension from its property, in the issue of any species.

It is indeed growen to be a great question, whether the transmutation of a species be possible either in the vegetable, Animal, or Minerall kingdome. For the possibility

possibility of it in the vegetable: I have heard Mr. *Bobart* and his *Son* often report it, and proffer to make oath that the Crocus and Gladiolus, as likwise the Leucoium, and Hyacinths by a long standing without replanting have in his garden changed from one kind to the other: and for satisfaction about the curiosity in the presence of Mr. *Boyle* I tooke up some bulbs of the very numericall roots whereof the relation was made, though the alteration was perfected before, where we saw the diverse bulbs growing as it were on the same stoole, close together, but no bulb half of the one kind, and the other half of the other: But the change-time being past it was reason we should beleive the report of good artists in matters of their own faculty.

Mr. *Wrench* a skilfull, and industrious gardiner for fruit and kitching-plants told me that the last year there was a change betwixt the kinds of the Coleflower, and the cabbage. Others I know who as from their experience most confidently affirme that they have prime-roses of the milk white colour, the root whereof before in another ground bare Oxelips: and it is usually beleived that divers single flowers may be changed into double by frequent transplantations, made into better grounds. I knew those that have had the wood Anemonies, and Colchiums double, who affirme that they took them into their garden wild, and single, and that that change was made by the soyle, and culture of the place.

For the animall Kingdome the instances of transmutation are in silkwormes, cadiz, and all caterpillars, which after a long sleep from the reptile turne into the volatile kind.

The minerall Kingdom is supposed to be famous and

and fruitfull in these changes, the hope of the Philosophers stone, or perfecting medicine requiring this beleife: Yet I am perswaded that in many of their changes they rather separate, and bring to apparence a latent minerall, than produce it by the transmutation of another into that nature. *Sennertus* recants those writtings of his, that affirmed iron to have been turned into copper by naturall, and artificiall waters of Vitrioll. The effect only in his second, and more mature judgment being the separation of a copper before latent in the Vitrioll, and the precipitation of it by the parts of the iron: and I have seen some experiments made by the honorable Person, for whom I am now writing, that have added strength to my former perswasion, particularly the supposed transmutation of quicksilver into lead, published as real by the learned *Vntzerus* and others, and to be made by dissolving the quicksilver in aqua fortis, & precipitating it by the tincture of Minium, proved but sophisticall, the Lead produced that way being indeed not made of the Mercury, but only reduced out of the tincture of Minium, wherein it lurck't, as that Gentleman doth more circumstantially set down in his own papers, and others of the like nature, which it were not proper here further to insist on.

It is a question, whether there be any reall transmutation, from the vegetable to the minerall kingdome, in petrifaction of any sort of wood: those petrifactions, which I have seen in England, are made thus, some particles of stone, that impregnate the body of water, make a crust about the stick that is to be petrified, and enter into the pores thereof, as fast as they are layed open by the water, washing through the stick, wherein there interceeds, noe change of the

the same parts, but by addition of some, and substraction of others, if I imagine aright, the new effect is wrought. The proof whereof may be, that the fibres of wood appear visible and to the touch and taste amidst the body of the stone.

In Ireland there is a Lake wherein (as that Noble Person I but now mentioned, hath related to me) there is soe great a petrifying faculty that the best whetstones used in that nation, are made of wood, cast therein to be petrified. In which stones though all the lineaments of the woody fibres remain, yet they are indued with the hardnesse, and other qualities of an exact stone. And Corall, the entire stonynesse thereof noe man can doubt, may well be imagined to be originally a vegetable bearing root, stalk, and leafe; and that afterward it is turned into its hardnesse by the peculiar property of the water: whether these operations of nature are likewise perfected by addition and substraction of parts only, or whether it be required that some parts for the production of this effect be transmuted I shall not determine.

And for the deciding the whole question, if the form be specificall, and so made by the aggregation of a certain number of accidents, those accidents & that number must be assigned that are thought enough to compleat a new form, before we may begin to judge in this matter for that very many accidents maybe changed it appears by the above named instances in vegetables & in other bodyes many more: Vinegerand Wine, are the same parts transposed and yet there seemes to be more difference between them than between Endive and Cichory, Maidenhaire and Scolopendrium, Rubarb and Dockes, which are in Vegetables esteemed for diverse species formally or specifically distinguished.

[*Editor's Note:* Material has been omitted at this point.]

3A

Reprinted by permission from pages 71–74 of *The Earthly Venus,* trans. S. B. Boas, Johnson Reprint Corporation, New York and London, 1966

PRODUCTION OF NEW SPECIES

P.-L. M. de Maupertuis

The theories above of eggs and worms may lend themselves too easily to the explanation of the origin of blacks and whites. They even try to explain how different species have come from the same individuals, but we saw in the preceding analysis what difficulties arise.

The varieties in the human species are not reduced to white or black but there are a thousand others. Those that strike us easily are probably not more difficult for Nature to create than others barely noticeable. If it were possible to understand the principle from decisive experiments, it might not seem stranger for a white child to be born from black parents than it would be for a blue-eyed child to come from a line of black-eyed ancestors.

Children usually resemble their parents and the variations among them at birth are often the results of various resemblances. Could we follow these variations back, we might find their origin in a common but unknown ancestor. They are kept alive by generations that have these traits and fade out with generations which do not have them. But what is even more surprising is to find such variations reappearing after having disappeared. We find a child who resembles neither his father nor his mother but has the features of his grandfather. Such facts, though amazing, are too frequent to be considered as doubtful.

Nature holds the source of all these varieties, but chance or art sets them going. So that people whose work is to satisfy the tastes of curiosity seekers become practically creators of new species. We find new breeds of dogs, pigeons, canaries appearing on the market, though they did not exist in nature. At first they were individual freaks, but art and repeated generations turned them into new species. The famed Lyonnés[40] creates each year a new variety and destroys the ones no longer in style. He corrects the shapes and varies the colors to the point of inventing species, such as the Harlequin Dane and the *Mopse* [Pug dog].

Why is this art restricted to animals? Why don't the bored Sultans in their seraglios, filled with women of all known races, have them bear new species? Were I reduced, as they are, to the only pleasure that form and features can give, I would soon have recourse to greater varieties. But, however beautiful the women born for them might be, they would know only the smallest share of love's pleasures as long as they remained ignorant of the pleasures of the mind and the heart.

Although we do not find among ourselves the creation of such new types of beauty, only too often do we see human beings who are of the same category for men of science, namely, the cross-eyed, the lame, the gouty, and the tubercular. Unfortunately, in order to fix their strain there is no need of a long series of generations. But wise Nature, because of the disgust she has inspired for these defects, has not desired that they be continued. Conse-

[40] Pierre Lyonnet, as the name is spelled in the various reference books, was born in Maestricht in 1707 and died in 1789. He was an artist and engraver as well as a naturalist, illustrating Lesser's "Théologie des insectes" (French translation in 1742). He also made the drawings for Trembley's "Mémoire pour servir à l'histoire d'un nouveau genre de polypes d'eau douce" ("Preliminary essay towards the natural history of a new genus"—or perhaps species—"of freshwater polyp").—A.

quently beauty is more apt to be hereditary. The slim waist and the leg that we admire are the achievements of many generations which have applied themselves to form them.

A Northern king was able to elevate and beautify his nation. His taste for men of height and fine faces was excessive and he induced them to come to his kingdom by various means. Fortune came to men whom Nature had made tall. Today we now see a singular example of the power of kings. This nation is distinguished for its tall men and regular features. So it is with a forest whose trees dominate all the neighboring woods, if the attentive eye of the master forester takes care to cultivate only trees that are straight and well chosen. The oak and the elm, adorned with the greenest of leaves, lift their branches to the sky where only the eagle can reach their crest. This king's successor today adorns the forest with laurel, myrtle, and flowers.

The Chinese once believed that the greatest beauty for women was to have feet so small that they could not stand on them. This nation, deeply attached to following the opinions and taste of its ancestors, finally reached the point of having women with ridiculous feet. I have seen Chinese women's mules into which our women could barely put a toe. This type of beauty is not new. Pliny, quoting Eudoxus, speaks of a people in India whose women have such small feet that they were called ostriches' feet.[41] It is true that he adds that men's feet measured a cubit and it might be that the smallness of the

[41] Pliny, "Natural History," Book VII, Chap. ii [Ed. Sillig, 1852, Vol. II, p. 9. Properly speaking, *struthopodes* should be translated "sparrow-footed" but Maupertuis was probably recalling the "great sparrows" of Xenophon.—A.]

women's feet led to exaggerating the size of the men's. Was not this nation, China, little known at the time? As a matter of fact the small size of Chinese women's feet cannot be attributed to nature, for during early childhood their feet are bound tightly to keep them from growing. In spite of this it would seem that Chinese women are born with smaller feet than women of other nations. This is a strange statement to make and it deserves attention from travelers.

Fatal beauty, the desire to please—what disorders do you not cause in the world! Not only do you torment our hearts, but you change nature's order. When a young French woman mocks the Chinese, she is not blaming her for thinking that she will be more attractive by sacrificing a graceful walk to the smallness of her feet. She knows at heart that this is not too high a price to pay for being charming, even if it means torture and pain. She herself since childhood has her body encased in a contraption of whalebone, or firmly held by an iron cross, both of which are more uncomfortable than all the bands tightened round the feet of the Chinese. Her head at night, bristling with curlers instead of being wrapped in the softness of her hair, can rest only on sharp points. In spite of this she sleeps peacefully, resting on her charms.

3B

Reprinted by permission from pages 78–80 of *The Earthly Venus,* trans. S. B. Boas, Johnson Reprint Corporation, New York and London, 1966

ATTEMPT AT AN EXPLANATION OF THE PRECEDING PHENOMENA

P.-L. M. de Maupertuis

In order to explain all the phenomena above: the accidental production of varieties, the succession of such varieties from one generation to the next, and finally the establishment of the destruction of these breeds, this, I believe, is what faces us. If what I am about to say is revolting to you, just consider it as an effort toward a satisfactory explanation. I do not expect it to be complete, for the phenomena are most complex. At least much shall be accomplished if I have been able to link these phenomena to others on which they are dependent.

We should take for granted facts that experience forces upon us:

(1) That seminal fluid from members of each species of animal contains a multitude of parts suitable to the formation of animals of the same species;

(2) That in the seminal fluid of each individual the parts suitable for the formation of features similar to those of that individual are normally in greater number and have a stronger affinity for one another. There are also many others which may form different features;

(3) As to the matter of the seminal fluid of each animal from which parts resembling it are to be formed, it would be an audacious guess, but might not be unlikely, to suggest that each part furnishes its germs. Experiments might throw light on this point. If it were tried for a long

time to mutilate some animals of the same breed, generation after generation, perhaps we might find the amputated parts diminishing little by little. Finally, one might see them disappear.[46]

The guesses above seem necessary and, once we admit them, it may seem possible to explain the many phenomena we have explored.

Parts similar to those of the father and mother, being the most numerous as well as having the greatest affinity, will be the ones to unite most easily and then will form animals like the ones from which they came.

Chance or a shortage of family traits will at times cause other combinations, and then we may see a white child born of black parents, or even a black child from white parents, though this is a much rarer phenomenon than the former.

I am speaking here only of these strange occurrences when a child born of parents of the same race has traits which he does not inherit from them, for we know that when the races are mixed, the child inherits from both.

The unusual unions of parts which are not the parts similar to those of the parents are really freaks to the bold who seek an explanation of Nature's wonders. For the wise man, satisfied with the spectacle, they are beauties.

To begin with, these latter productions are accidental, for the parts whose origin is ancestral are the more abundant in the seeds. Therefore after a few generations, or even in the next generation, the original species will regain its strength and the child, instead of resembling its father or

[46] It may perhaps be superfluous to recall the experiment of Weismann here, since it is well known. But at least it is worth pointing out that it was anticipated by Maupertuis.—A.

mother, will resemble some distant ancestor.[47] In order to create species from races that become established, it is really necessary to have the same types unite for several generations. The parts suitable to recreate the original trait, since they are less numerous in each generation, are either lost or remain in so small a quantity that a new chance event would be needed to reproduce the original species. However, though I imagine the basic stock of all these varieties is to be found in the seminal fluids themselves, I do not exclude the possible influence of climate and food. It would seem that the heat of the Torrid Zone is more favorable to the particles that compose black skin than to those that make up white skin. And I simply do not know how far this kind of influence of climate and food may go after many centuries.

It would indeed be something to occupy the attention of Philosophers if they would try to discover whether certain unnatural characters induced in animals for many generations were transmitted to their descendants. Whether tails or ears cut off from generation to generation did diminish in size or even finally disappear would be of importance.

One thing is certain, and that is that all variations which may characterize new breeds of animals and plants tend to degenerate. They are the sports of Nature, only preserved through art or discipline, for her original creations always tend to return.

[47] This happens daily in families. A child who resembles neither his father nor his mother will resemble a grandparent.

4

Reprinted from pages 110–117 of *The Farmer's Tour Through the East of England. Being the Register of a Journey Through the Various Counties of This Kingdom, to Enquire into the State of Agriculture, &c.*, vol. I, London, 1771

THE FARMER'S TOUR THROUGH THE EAST OF ENGLAND

A. Young

[*Editor's Note:* Material has been omitted at this point.]

Mr. *Bakewell* of *Diſhley*, one of the moſt conſiderable farmers in this country, has in ſo many inſtances improved on the huſbandry of his neighbours, that he merits particular notice in this journal.

His breed of cattle is famous throughout the kingdom; and he has lately ſent many to *Ireland*. He has in this part of his buſineſs many ideas which I believe are perfectly new; or that have hitherto been totally neglected. This principle is to gain the beaſt, whether ſheep or cow, that will weigh moſt in the moſt valuable joints:— there is a great difference between an ox of 50 ſtone, carrying 30 in roaſting pieces, and 20 in coarſe boiling ones— —and another carrying 30 in the latter, and 20 in the former. And at the ſame time that he gains the ſhape, that is, of the greateſt value in the ſmalleſt compaſs; he aſſerts, from long experience, that he gains a breed

much

The prints of this material were made from microfilm that was made available through the courtesy of the History of Sciences Collections, University of Oklahoma Libraries.

much hardier, and eaſier fed than any others. Theſe ideas he applies equally to ſheep and oxen.

In the breed of the latter, the old notion was, that where you had much and large bones, there was plenty of room to lay fleſh on; and accordingly the graziers were eager to buy the largeſt boned cattle. This whole ſyſtem Mr. *Bakewell* has proved to be an utter miſtake. He aſſerts, the ſmaller the bones, the truer will be the make of the beaſt——the quicker ſhe will fat——and her weight, we may eaſily conceive, will have a larger proportion of valuable meat: *fleſh*, not *bone*, is the butcher's object. Mr. *Bakewell* admits that a large boned beaſt, may be made a large fat beaſt, and that he may come to a great weight; but juſtly obſerves, that this is no part of the profitable enquiry; for ſtating ſuch a ſimple propoſition, without at the ſame time ſhewing the expence of covering thoſe bones with fleſh, is offering no ſatisfactory argument. The only object of real importance, is the proportion of *graſs* to *value*. I have 20 acres; which will pay me for thoſe acres beſt, large or ſmall boned cattle? The lat-

ter

ter fat ſo much quicker, and more profitably in the joints of value; that the query is anſwered in their favour from long and attentive experience.

Among other breeds of cattle the *Lincolnſhire* and the *Holderneſs* are very large, but their ſize lies in their bones: they may be fattened to great loſs to the grazier, nor can they ever return ſo much for a given quantity of graſs, as the ſmall boned, long horned kind.

The breed which Mr. *Bakewell* has fixed on as the beſt in *England*, is the *Lancaſhire*, and he thinks he has improved it much, in bringing the carcaſs of the beaſt into a truer mould; and particularly by making them broader over the backs. The ſhape which ſhould be the criterion of a cow, a bull, or an ox, and alſo of a ſheep, is that of an hogſhead, or a firkin; truly circular with ſmall and as ſhort legs as poſſible: upon the plain principle, that the value lies in the barrel, not in the legs. All breeds, the backs of which riſe in the leaſt ridge, are bad. I meaſured two or three cows, 2 feet 3 inches flat acroſs their back from hip to hip—and their legs remarkably ſhort.

Mr.

Mr. *Bakewell* has now a bull of his own breed which he calls *Twopenny*, which leaps cows at 5*l.* 5*s.* a cow. This is carrying the breed of horned cattle to wonderful perfection. He is a very fine bull—most truly made, according to the principles laid down above. He has many others got by him, which he lets for the season, from 5 guineas to 30 guineas a season, but rarely sells any. He would not take 200*l.* for *Twopenny*. He has several cows which he keeps for breeding, that he would not sell at 30 guineas apiece.

Another particularity is the amazing gentleness in which he brings up these animals. All his bulls stand still in the field to be examined: the way of driving them from one field to another, or home, is by a little swish; he or his men walk by their side, and guide him with the stick wherever they please; and they are accustomed to this method from being calves. A lad, with a stick three feet long, and as big as his finger, will conduct a bull away from other bulls, and his cows from one end of the farm to the other.

All this gentleneſs is merely the effect of management, and the miſchief often done by bulls, is undoubtedly owing to practices very contrary—or elſe to a total neglect.

The general order in which Mr. *Bakewell* keeps his cattle is pleaſing; all are fat as bears; and this is a circumſtance which he inſiſts is owing to the excellence of the breed. His land is no better than his neighbours, at the ſame time that it carries a far greater proportion of ſtock; as I ſhall ſhew by and by. The ſmall quantity, and the inferior quality of food that will keep a beaſt perfectly well made, in good order, is ſurprizing: ſuch an animal will grow fat in the ſame paſture that would ſtarve an ill made, great boned one.

In the breed of his ſheep, Mr. *Bakewell* is as curious, and I think, if any difference, with greater ſucceſs, than in his horned cattle: for better made animals cannot be ſeen than his rams and ewes: their bodies are as true barrels as can be ſeen *; round, broad

* The following is an account of two ſheep of Mr. *Bakewell*'s, meaſured in the wool.

" I this

broad backs; and the legs not above six inches long: and a most unusual proof of kindly fattening, is their feeling quite fat, just within their fore legs on the ribs, a point in which sheep are never examined in common; from common breeds never carrying any fat there.

In his breed of sheep, he proceeds exactly on the same principle as with oxen; the fatting in the valuable parts of the body;

" I this day measured Mr. *Bakewell's* three years old ram, and found him as follows:

	Feet.	*Inches.*
His girt, - - - - -	5	10
His height, - - - - -	2	5
His collar broad at ear tips, -	1	4
Broad over his shoulders, -	1	11 ½
Ditto over his ribs, - -	1	10 ½
Ditto his hips, - - -	1	9 ½

Dishley, 17th *March*, 1770.

H. SANDFORD.

" This day measured a two year old barren ewe.

	Feet.	*Inches.*
Height, - - - - -	1	11
Girt, - - - - - -	5	9

Breast from the ground, the breadth of 4 fingers.

N. B. I would have measured her breadth, but for a fall of snow.

Dishley, ut sup.

H. S.

I 2 and

and the living on much poorer food than other ſorts. He has found from various experience in many parts of the kingdom, as well as upon his own farm, that no land is too bad for a *good* breed of cattle, and particularly ſheep. It may not be proper for large ſtock, that is large boned ſtock, but undoubtedly more proper for a valuable well made ſheep than the uſual wretched ſorts found in moſt parts of *England* on poor ſoils—ſuch as the moor ſheep—the *Welch* ones—and the *Norfolks.*—And he would hazard any moderate ſtake, that his own breed, each ſheep of which is worth ſeveral of thoſe poor ſorts, would do better on thoſe poor ſoils than the ſtock generally found on them: A good and true ſhape having been found the ſtrongeſt indication of hardineſs, and what the graziers call a *kindly* ſheep; one that has always an inclination to feed.

He has an experiment to prove the hardineſs of his breed which deſerves notice. He has 5 or 6 ewes, that have gone conſtantly in the highways ſince *May-day*, and have never been in his fields: the roads are

narrow,

narrow, and the food very bare; they are in excellent order, and nearly fat; which proves in the ſtrongeſt manner, the excellence of the breed. And another circumſtance of a peculiar nature is his flock of ewes, that have reared two lambs, being quite fat in the firſt week of *July*; an inſtance hardly to be paralleled.

The breed is originally *Lincolnſhire*, but Mr. *Bakewell* thinks, and very juſtly, that he has much improved it. The grand profit, as I before obſerved, is from the ſame food going ſo much farther in feeding theſe than any others; not however that Mr. *Bakewell*'s breed is ſmall; on the contrary, it is as weighty as nine tenths of the kingdom; for he ſells fat wethers at three years and an half old at 2 *l.* a head. Other collateral circumſtances of importance are the wool being equal to any other; and the ſheep ſtanding the fold better. He ſells no tups, but lets them at from 5 guineas to 30 guineas for the ſeaſon.

[*Editor's Note:* Material has been omitted at this point.]

5

Reprinted from pages 363–366 of *Communications to the Board of Agriculture; on Subjects Relative to the Husbandry and Internal Improvement of the Country,* vol. 1, parts III and IV, London, 1797

LETTER FROM DR. PRIESTLEY TO SIR JOHN SINCLAIR, PRESIDENT OF THE BOARD OF AGRICULTURE

J. Priestley

SIR, Philadelphia, April 9th, 1797.

THOUGH not employed in Agriculture, and my philosophical pursuits have had other objects, I have not been wholly inattentive to a subject of so much importance; and though I am not able to supply you with any thing out of my own stores, I am happy to have it in my power to communicate something from the labours and observations of others.

I have fortunately become acquainted with Mr. Joseph Cooper, who lives opposite to this city, on the Jersey shore, a great original genius in agriculture, and farming in general. Without any advantage of education superior to other farmers, he has thought philosophically on the subject, and has had very extraordinary success, in a variety of plans which are wholly new, and which promise to be of great benefit to his country and the world.

I have his leave to communicate to you his observations and experiments relating to an opinion and practice which has prevailed, I believe universally, but which he is satisfied is ill founded. Plants, it is said, will degenerate, unless the soil in which they grow be changed. It is therefore thought to be necessary from time to time to get fresh seeds and roots, &c. from distant places. Mr. Cooper, on the contrary, has for many years been in the habit of selecting the best seeds and roots of his own, and though he has continually sown and planted them in the same soil, every article of his produce is greatly superior to those of any other person who supplies this market, and they seem to be still in a state of improvement. This, without his knowing it, is the very same plan that was adopted by Mr. Bakewell in England, with respect to animals. He kept improving his breeds, by only coupling those in which the properties he wished to produce were the most conspicuous, without any regard to consanguinity, or any other circumstance whatever.

Mr. Cooper was led to his present practice, which he began more than forty years ago, by observing that vegetables of all kinds were very subject to change with respect to their time of coming to maturity, and other properties, but that the best seeds never failed to produce the best plants. Among a great number of experiments he particularly mentions the following:

About the year 1746, his father procured seeds of the long watery squash, and though they have been used on the farm ever since that time, without any change, they are at this time better than they were at the first.

His early peas were procured from London in the year 1756; and though they have been planted on the same place every season, they have been so far from degenerating, that they are preferable to what they were then. The seeds of his asparagus he had from New York in 1752, and though they have been treated in the same manner, the plants are greatly improved.

It is more particularly complained, that potatoes degenerate when they are planted from the same roots in the same place. At this Mr. Cooper says he does not wonder, when it is customary with farmers to use the best, and plant from the refuse; whereas, having observed that some of his plants produced potatoes that were larger, better shaped, and in greater abundance than others, he took his seed from them only; and the next season he found that the produce was of a quality superior to any that he had ever had before. This practice he still continues, and finds that he is abundantly rewarded for his trouble.

Mr. Cooper is also careful to sow the plants, from which he raises his seed, at a considerable distance from any others. Thus, when his radishes are fit for use, he takes ten or twelve that he most approves, and plants them at least one hundred yards from others that blossom at the same time. In the same manner he treats all his other plants, varying the circumstances according to their nature.

About the year 1772, a friend of his sent him a few grains of a small kind of Indian corn, not larger than goose-shot, which produced from eight to ten ears on a stalk. They were also small, and he found that few of them ripened before the frost. Some of the largest and earliest of these he saved, and planting them between rows of a larger and earlier kind, the produce was much improved. He then planted from those that had produced the greatest number of the largest ears, and that were the first ripe; and the next season the produce, with respect to quality and quantity, was preferable to any that he had ever planted before. From this corn he has continued to plant ever since, selecting his seed in the following manner:

When the first ears are ripe enough for seed, he gathers a sufficient quantity for early corn, or for replanting, and at the time that he wishes his corn to be generally ripe, he gathers a sufficient quantity for the next year's planting; having particular care to take it from stalks that are large at the bottom, of a regular taper, not very tall, the ears set low, and containing the greatest number of good sizable ears, and of the best quality; these he dries quickly, and from them he plants his main crop; and if any hills be missing, he replants from the seeds that were first gathered, which he says will cause the crops to ripen more regularly than they commonly do, and which is of great advantage. This method he has practiced many years, and he is satisfied that it has been the means of increasing the quantity, and improving the quality, of his crops beyond what any person who had not tried the experiment could imagine.

Farmers differ much with respect to the distance at which they plant their corn, and the number of grains they put in a hill. Different soils, Mr. Cooper observes, may require different practices in both these respects; but in every kind of soil that he has tried, he finds that planting the rows six feet asunder each way, as nearly at right angles as may be, and leaving not more than four stalks in a hill, produces the best crop. The common method of saving seed-corn, by taking the ears from the heap, is attended, he says, with two disadvantages; one is the taking the largest ears, of which in general only one grows on a stalk, which lessens the produce; and the other is taking ears that ripen at different times.

For many years Mr. Cooper renewed all the seed of his winter grain from a single plant, which he had observed to be more productive, and of a better quality than the rest, which he is satisfied has been of great use. And he is of opinion, that all kinds of garden vegetables may be improved by the methods described above, particular care being taken that different kinds of the same vegetables do not bloom at the same time near together, since by this means they injure one another.

It is alleged, that foreign flax seed produces the best flax in Ireland; but Mr. Cooper says, that when it is considered that only the bark of the plant is used, and that this is in perfection before the seed is ripe, it will appear that his hypothesis is not affected by it.

Mr. Cooper had the following instance of the naturalization of a plant in a different climate: he had some water-melon seed sent to him from Georgia, which he was informed was of a peculiarly good quality; knowing that seeds from vegetables which grow in a hot climate require a longer summer than that of Pennsylvania, he gave them the most favourable situation that he had, and used glasses to forward their growth, and yet few of them ripened well. But finding them to be of an excellent quality, he saved the seeds of those that ripened the first; and by continuing this practice five or six years, they came to ripen as early as any that he ever had.

I cannot express how much I admire the exertions already made with respect to the great objects pursued by the Board of Agriculture. They promise to counteract the destructive effects of war, and in time of peace will, I hope, speedily repair all the calamities occasioned by it, as it can be done by the better condition of those who survive them. I particularly admire the liberality of your Address to all nations, on a subject so highly interesting to them all; and I promise myself a new and more happy era in the state of society from it. With the greatest respect, I am, Sir,

Yours sincerely,

J. PRIESTLEY.

P. S. I am directing a few experiments on the use of gypsum as a manure, which I think will ascertain the principle on which it acts, and may lead to a more effectual application of it. If I have any success, you shall hear from me again.

6

Reprinted from pages 163–171 of *An Essay on the Principle of Population, as It Affects the Future Improvement of Society. With Remarks on the Speculations of Mr. Godwin, M. Condorcet and Other Writers,* J. Johnson, London, 1798

AN ESSAY ON THE PRINCIPLE OF POPULATION

T. R. Malthus

[*Editor's Note:* In the original, material precedes and follows this excerpt.]

which the argument rests, is, that because the limit of human life is undefined ; because you cannot mark its precise term, and say so far exactly shall it go and no further ; that therefore its extent may increase for ever, and be properly termed, indefinite or unlimited. But the fallacy and absurdity of this argument will sufficiently appear from a slight examination of what Mr. Condorcet calls the organic perfectibility, or degeneration, of the race of plants and animals, which he says may be regarded as one of the general laws of nature.

I am told that it is a maxim among the improvers of cattle, that you may breed to any degree of nicety you please, and they found this maxim upon another, which is, that some of the offspring will possess the desirable qualities of the parents in a greater degree. In the

famous

famous Leicestershire breed of sheep, the object is to procure them with small heads and small legs. Proceeding upon these breeding maxims, it is evident, that we might go on till the heads and legs were evanescent quantities ; but this is so palpable an absurdity, that we may be quite sure that the premises are not just, and that there really is a limit, though we cannot see it, or say exactly where it is. In this case, the point of the greatest degree of improvement, or the smallest size of the head and legs, may be said to be undefined, but this is very different from unlimited, or from indefinite, in Mr. Condorcet's acceptation of the term. Though I may not be able, in the present instance, to mark the limit, at which further improvement will stop, I can very easily mention a point at which it will not arrive. I should not scruple to assert, that were the breeding

to

to continue for ever, the head and legs of these sheep would never be so small as the head and legs of a rat.

It cannot be true, therefore, that among animals, some of the offspring will possess the desirable qualities of the parents in a greater degree ; or that animals are indefinitely perfectible.

The progress of a wild plant, to a beautiful garden flower, is perhaps more marked and striking, than any thing that takes place among animals, yet even here, it would be the height of absurdity to assert, that the progress was unlimited or indefinite. One of the most obvious features of the improvement is the increase of size. The flower has grown gradually larger by cultivation. If the progress were really unlimited, it might be increased ad infinitum ; but

this

this is so gross an absurdity, that we may be quite sure, that among plants, as well as among animals, there is a limit to improvement, though we do not exactly know where it is. It is probable that the gardeners who contend for flower prizes have often applied stronger dressing without success. At the same time, it would be highly presumptuous in any man to say, that he had seen the finest carnation or anemone that could ever be made to grow. He might however assert without the smallest chance of being contradicted by a future fact, that no carnation or anemone could ever by cultivation be increased to the size of a large cabbage ; and yet there are assignable quantities much greater than a cabbage. No man can say that he has seen the largest ear of wheat, or the largest oak that could ever grow ; but he might easily, and with perfect certainty, name a point

point of magnitude, at which they would not arrive. In all these cases therefore, a careful distinction should be made, between an unlimited progress, and a progress where the limit is merely undefined.

It will be said, perhaps, that the reason why plants and animals cannot increase indefinitely in size, is, that they would fall by their own weight. I answer, how do we know this but from experience? from experience of the degree of strength with which these bodies are formed. I know that a carnation, long before it reached the size of a cabbage, would not be supported by its stalk; but I only know this from my experience of the weakness, and want of tenacity in the materials of a carnation stalk. There are many substances in nature

ture of the same size that would support as large a head as a cabbage.

The reasons of the mortality of plants are at present perfectly unknown to us. No man can say why such a plant is annual, another biennial, and another endures for ages. The whole affair in all these cases, in plants, animals, and in the human race, is an affair of experience; and I only conclude that man is mortal, because the invariable experience of all ages has proved the mortality of those materials of which his visible body is made.

"What can we reason but from what we know."

Sound philosophy will not authorize me to alter this opinion of the mortality of man on earth, till it can be clearly proved, that the human race has made,

made, and is making, a decided progress towards an illimitable extent of life. And the chief reason why I adduced the two particular instances from animals and plants, was to expose, and illustrate, if I could, the fallacy of that argument, which infers an unlimited progress, merely because some partial improvement has taken place, and that the limit of this improvement cannot be precisely ascertained.

The capacity of improvement in plants and animals, to a certain degree, no person can possibly doubt. A clear and decided progress has already been made; and yet, I think it appears, that it would be highly absurd to say, that this progress has no limits. In human life, though there are great variations from different causes, it may be doubted, whether, since the world began, any organic

organic improvement whatever in the human frame can be clearly ascertained. The foundations therefore, on which the arguments for the organic perfectibility of man rest, are unusually weak, and can only be considered as mere conjectures. It does not, however, by any means, seem impossible, that by an attention to breed, a certain degree of improvement, similar to that among animals, might take place among men. Whether intellect could be communicated may be a matter of doubt: but size, strength, beauty, complexion, and perhaps even longevity are in a degree transmissible. The error does not seem to lie, in supposing a small degree of improvement possible, but in not discriminating between a small improvement, the limit of which is undefined, and an improvement really unlimited. As the human race however could not be improved in

this

this way, without condemning all the bad specimens to celibacy, it is not probable, that an attention to breed should ever become general ; indeed, I know of no well-directed attempts of the kind, except in the ancient family of the Bickerstaffs, who are said to have been very successful in whitening the skins, and increasing the height of their race by prudent marriages, particularly by that very judicious cross with Maud, the milk-maid, by which some capital defects in the constitutions of the family were corrected.

It will not be necessary, I think, in order more completely to shew the improbability of any approach in man towards immortality on earth, to urge the very great additional weight that an increase in the duration of life would give to the argument of population.

7

Reprinted from *R. Soc. (London) Philos. Trans.* **89**:195–204 (1799)

AN ACCOUNT OF SOME EXPERIMENTS ON THE FECUNDATION OF VEGETABLES

T. A. Knight

Read May 9, 1799.

Elton, April 25, 1799.

THE result of some experiments which I have amused myself in making on plants, appearing to me to be interesting to the naturalist, by proving the existence of superfœtation in the vegetable world, and being likely to conduce to some improvements in agriculture, I have taken the liberty to communicate them to you.

The breeders of animals have very long entertained an opinion, that considerable advantages are obtained by breeding from males and females not related to each other. Though this opinion has lately been controverted, the number of its opposers has gradually diminished; and I can speak from my own observation and experience, that animals degenerate, in size at least, on the same pasture, and in other respects under the same management, when this process of crossing the breed is neglected.

The close analogy between the animal and vegetable world, and the sexual system equally pervading both, induced me to suppose, that similar means might be productive of similar effects in each; and the event has, I think, fully justified this opinion. The principal object I had in view, was to obtain new and improved varieties of the apple, to supply the place of those which have become diseased and unproductive, by having been cultivated beyond the period which nature appears to have assigned to their existence. But, as I foresaw that several years must elapse, before the success or failure of this process could possibly be ascertained, I wished, in the interval, to see what would be its effects on annual plants. Amongst these, none appeared so well calculated to answer my purpose as the common pea; not only because I could obtain many varieties of this plant, of different forms, sizes, and colours; but also, because the structure of its blossom, by preventing the ingress of insects and adventitious farina, has rendered its varieties remarkably permanent. I had a kind growing in my garden, which, having been long cultivated in the same soil, had ceased to be productive, and did not appear to recover the whole of its former vigour, when removed to a soil of a somewhat different quality; on this, my first experiment, in 1787, was made. Having opened a dozen of its immature blossoms, I destroyed the male parts, taking great care not to injure the female ones; and, a few days afterwards, when the blossoms appeared mature, I introduced the farina of a very large and luxuriant gray pea into one half of the blossoms, leaving the other half as they were. The pods of each grew equally well; but I soon perceived, that in those into whose blossoms the farina had not been introduced, the seeds remained nearly as they were before the blossoms expanded, and in that state they withered. Those in the other pods attained maturity, but were not in any sensible degree different from those afforded by other plants of the same variety; owing, I imagine, to the external covering of the seed (as I have found in other plants) being furnished entirely by the female. In the succeeding spring, the difference, however, became extremely obvious; for the plants from them rose with excessive luxuriance, and the colour of their leaves and stems clearly indicated, that they had all exchanged their whiteness for the colour of the male parent: the seeds produced in autumn were dark gray. By introducing the farina of another white variety, (or, in some instances, by simple culture,) I found this colour was easily discharged, and a numerous variety of new kinds produced, many of which were, in size, and in every other respect, much superior to the original white kind, and grew with excessive luxuriance, some of them attaining the height of more than twelve feet. I had frequent occasion to observe, in this plant, a stronger tendency to produce purple blossoms; and coloured seeds, than white ones; for, when I introduced the farina of a purple blossom into a white one, the whole of the seeds in the succeeding year became coloured; but, when I endeavoured to discharge this colour, by reversing the process, a part only of them afforded plants with white blossoms; this part sometimes occupying one end of the pod, and being at other times irregularly intermixed with those which, when sown, retained their colour. It may perhaps be supposed, that something might depend on the quantity of farina employed; but I never could discover, in this, or in any other experiment, in which superfœtation did not take place, that the largest or smallest quantity of farina afforded any difference in the effect produced.

The dissimilarity I observed in the offspring afforded by different kinds of farina, in these experiments, pointed out to me an easy method of ascertaining whether superfœtation (the existence of which has been admitted amongst animals) could also take place in the vegetable world. For, as the offspring of a white pea is always white, unless the farina of a coloured kind be introduced into the blossom, and, as the colour of the gray one is always transferred to its offspring, though the female be white, it readily occurred to me, that if the farina of both were mingled, or applied at the same moment, the offspring of each could be easily distinguished.

My first experiment was not altogether successful; for the offspring of five pods (the whole which escaped the birds) received their colour from the coloured male. There was, however, a strong resemblance to the other male, in the growth and character of more than one of the plants; and the seeds of several, in the autumn, very closely resembled it in every thing but colour. In this experiment, I used the farina of a white pea, which possessed the remarkable property of shrivelling excessively when ripe; and, in the second year, I obtained white seeds, from the gray ones above mentioned, perfectly similar to it. I am strongly disposed to believe, that the seeds were here of common parentage; but I do not conceive myself to be in possession of facts sufficient to enable me to speak with decision on this question.

If, however, the female afford the first organised atom, and the farina act only as a stimulus, it appears to me by no means impossible, that the explosion of two vesicles of farina, at the same moment, (taken from different plants,) may afford seeds (as I have supposed) of common parentage; and, as I am unable to discover any source of inaccuracy in this experiment, I must believe this to have happened.

Another species of superfœtation (if I have justly applied

that term to a process in which one seed appears to have been the offspring of two males) has occurred to me so often, as to remove all possibility of doubt as to its existence. In 1797, the year after I had seen the result of the last mentioned experiment, having prepared a great many white blossoms, I introduced the farina of a white and that of a gray pea, nearly at the same moment, into each; and as, in the last year, the character of the coloured male had prevailed, I used its farina more sparingly than that of the white one; and now almost every pod afforded plants of different colours. The majority, however, were white; but the characters of the two kinds were not sufficiently distinct to allow me to judge with precision, whether any of the seeds produced were of common parentage or not. In the last year, I was more fortunate: having prepared blossoms of the little early frame pea, I introduced its own farina, and immediately afterwards that of a very large and late gray kind, and I sowed the seeds thus obtained in the end of the last summer. Many of them retained the colour and character of the small early pea, not in the slightest degree altered, and blossomed before they were eighteen inches high; whilst others, (taken from the same pods,) whose colour was changed, grew to the height of more than four feet, and were killed by the frost, before any blossoms appeared.

It is evident, that in these instances superfœtation took place; and it is equally evident, that the seeds were not all of common parentage. Should subsequent experience evince, that a single plant may be the offspring of two males, the analogy between animal and vegetable nature may induce some curious conjecture, relative to the process of generation in the animal world.

In the course of the preceding experiments, I could never

observe that the character, either of the male or female, in this plant, at all preponderated in the offspring; but, as this point appeared interesting, I made a few trials to ascertain it. And, as the foregoing observations had occurred in experiments made principally to obtain new and improved varieties of the pea, for garden culture, I chose, for a similar purpose, the more hardy varieties usually sown in the fields. By introducing the farina of the largest and most luxuriant kinds into the blossoms of the most diminutive, and by reversing this process, I found that the powers of the male and female, in their effects on the offspring, are exactly equal. The vigour of the growth, the size of the seeds produced, and the season of maturity, were the same, though the one was a very early, and the other a late variety. I had, in this experiment, a striking instance of the stimulative effects of crossing the breeds; for the smallest variety, whose height rarely exceeded two feet, was increased to six feet; whilst the height of the large and luxuriant kind was very little diminished. By this process, it is evident, that any number of new varieties may be obtained; and it is highly probable, that many of these will be found better calculated to correct the defects of different soils and situations, than any we have at present; for, I imagine that all we now possess, have in a great measure been the produce of accident; and it will rarely happen, in this or any other case, that accident has done all that art will be found able to accomplish.

The success of my endeavours to produce improved varieties of the pea, induced me to try some experiments on wheat; but these did not succeed to my expectations. I readily obtained as many varieties as I wished, by merely sowing the different kinds together; for the structure of the blossom of this plant (unlike that of the pea) freely admits the ingress of adventitious farina, and is thence very liable to sport in varieties. Some of those I obtained were excellent; others very bad; and none of them permanent. By separating the best varieties, a most abundant crop was produced; but its quality was not quite equal to the quantity, and all the discarded varieties again made their appearance. It appeared to me an extraordinary circumstance, that, in the years 1795 and 1796, when almost the whole crop of corn in the island was blighted, the varieties thus obtained, and these only, escaped, in this neighbourhood, though sown in several different soils and situations.

My success on the apple (as far as long experience and attention have enabled me to judge from the cultivated appearance of trees which have not yet borne fruit) has been fully equal to my hopes. But, as the improvement of this fruit was the first object of my attention, no probable means of improvement, either from soil or aspect, were neglected. The plants, however, which I obtained from my efforts to unite the good qualities of two kinds of apple, seem to possess the greatest health and luxuriance of growth, as well as the most promising appearance in other respects. In some of these, the character of the male appears to prevail; in others, that of the female; and in others, both appear blended, or neither is distinguishable. These variations, which were often observable in the seeds taken from a single apple, evidently arise from the want of permanence in the character of this fruit, when raised from seed.

The results of similar experiments on another fruit, the grape, were nearly the same as of those on the apple, except that, by mingling the farina of a black and a white grape, just as the blossoms of the latter were expanding, I sometimes obtained plants, from the same berry, so dissimilar, that I had good reason to believe them the produce of superfœtation. By taking off the cups, and destroying the immature male parts, (as in the pea,) I perfectly succeeded in combining the characters of different varieties of this fruit, as far as the changes of form, and autumnal tints, in the leaves of the offspring, will allow me to judge.

Many experiments, of the same kind, were tried on other plants; but it is sufficient to say, that all tended to evince, that improved varieties of every fruit and esculent plant may be obtained by this process, and that nature intended that a sexual intercourse should take place between neighbouring plants of the same species. The probability of this will, I think, be apparent, when we take a view of the variety of methods which nature has taken to disperse the farina, even of those plants in which it has placed the male and female parts within the same empalement. It is often scattered by an elastic exertion of the filaments which support it, on the first opening of the blossom; and its excessive lightness renders it capable of being carried to a great distance by the wind. Its position within the blossom, is generally well adapted to place it on the bodies of insects; and the villous coat of the numerous family of bees, is not less well calculated to carry it. I have frequently observed, with great pleasure, the dispersion of the farina of some of the grasses, when the sun had just risen in a dewy morning. It seemed to be impelled from the plant with considerable force; and, being blue, was easily visible, and very strongly resembled, in appearance, the explosion of a grain of gunpowder. An examination of the structure of the blossoms of many plants, will immediately point out, that nature has something more in view, than that its own proper males should fecundate each blossom; for the means it employs are always those best calculated to answer the intended purpose. But the farina is often so placed, that it can never reach the summit of the pointal, unless by adventitious means; and many trials have convinced me, that it has no action on any other part of it. In promoting this sexual intercourse between neighbouring plants of the same species, nature appears to me to have an important purpose in view; for, independent of its stimulative power, this intercourse certainly tends to confine within more narrow limits, those variations which accidental richness or poverty of soil usually produces. It may be objected, by those who admit the existence of vegetable mules, that, under this extensive intercourse, these must have been more numerous; but my total want of success, in many endeavours, to produce a single mule plant, makes me much disposed to believe that hybrid plants have been mistaken for mules; and to doubt (with all the deference I feel for the opinions of LINNÆUS and his illustrious followers) whether nature ever did, or ever will, permit the production of such a monster. The existence of numerous mules in the animal world, between kindred species, is allowed; but nature has here guarded against their production, by impelling every animal to seek its proper mate; and, amongst the feathered tribe, when, from perversion of appetite, sexual intercourse takes place between those of distinct genera,* it has, in some instances at least, rendered the death of the female the inevitable consequence. But, in the vegetable world, there is not any thing to direct the male to its proper female: its farina is carried, by winds and insects, to plants of every different

* This is said to be the case with the drake and the hen.

genus and species; and it therefore appears to me, (as vegetable mules certainly are not common,) that nature has not permitted them to exist at all.

I cannot dismiss this subject, without expressing my regret, that those who have made the science of botany their study, should have considered the improvement of those vegetables which, in their cultivated state, afford the largest portion of subsistence to mankind and other animals, as little connected with the object of their pursuit. Hence it has happened, that whilst much attention has been paid to the improvement of every species of useful animal, the most valuable esculent plants have been almost wholly neglected. But, when the extent of the benefit which would arise to the agriculture of the country, from the possession of varieties of plants which, with the same extent of soil and labour, would afford even a small increase of produce, is considered, this subject appears of no inconsiderable importance. The improvement of animals is attended with much expence, and the improved kinds necessarily extend themselves slowly; but a single bushel of improved wheat or peas, may in ten years be made to afford seed enough to supply the whole island; and a single apple, or other fruit-tree, may within the same time be extended to every garden in it. These considerations have been the cause of my addressing the foregoing observations to you at this time; for it was much my wish to have ascertained, before I wrote to you, whether in any instance a single plant can be the offspring of two male parents. The decision of that question must of necessity have occupied two years, and must therefore be left to the test of future experiment.

8

Reprinted from pages 448–451 of *Lectures on Comparative Anatomy, Physiology, Zoology, and the Natural History of Man*, R. Carlile, London, 1823

ORIGIN AND TRANSMISSION OF VARIETIES IN FORM

W. Lawrence

[*Editor's Note:* Material has been omitted at this point.]

The formation of new varieties by breeding from individuals in whom the desireable properties exist in the greatest degree, is seen much more distinctly in our domestic animals than in our own species, since the former are entirely in our power. The great object is to preserve the race pure, by selecting for propagation the animals most conspicuous for the size, colour, form, proportion, or any other property we may fix on, and excluding all others. In this way we may gain sheep valuable for their fleece, or for their carcase, large or small, with thick or thin legs; just such, in short, as we choose within certain limits.

The importance of this principle is fully understood in rearing horses. The Arabian preserves the pedigree of his horse more carefully than his own; and never allows any ignoble blood to be mixed with that of his valued breeds: he attests their unsullied nobility by formal depositions and numerous witnesses[59]. The English breeder knows equally well that he must vary his stallions and mares according as he wishes for a cart-horse, a riding-horse, or a racer; and that a mistake in this point would immediately frustrate his views. The distinguished and various excellencies, which the several English races of these useful animals have acquired, shew what close attention and perseverance can accomplish in the improvement of breed.

Blood is equally important in the cock; and the introduction of an inferior individual would inevitably deteriorate the properties of the offspring.

The hereditary transmission of physical and moral qualities, so well understood and familiarly acted on in the domestic animals, is equally true of man. A superior breed of human beings could only be produced by selections and exclusions similar to those so successfully employed in rearing our more

[59] "Several things concur, to maintain this perfection in the horses of Arabia, such, as the great care the Arabs take in preserving the breed genuine, and by permitting none but stallions of the first form to have access to the mares: this is never done but in the presence of a witness, the secretary of the emir, or some public officer; he attests the fact, records, the name of the horse, mare, and whole pedigree of each; and these attestations are carefully preserved, for on them depends the future price of the foal."

A copy of a public legal certificate given to the purchaser of an Arabian horse is added in a note. PENNANT'S *British Zoology*, v. ii. appendix 1.

Equal attention is paid to the breed of horses by the Circassians, who distinguish the various races by marks on the buttock. To imprint the character of noble descent on a horse of common race is a kind of forgery punished with death. PALLAS, *Travels in the Southern Provinces of the Russian Empire*; ch. xiv.

valuable animals. Yet, in the human species, where the object is of such consequence, the principle is almost entirely overlooked. Hence all the native deformities of mind and body, which spring up so plentifully in our artificial mode of life, are handed down to posterity, and tend by their multiplication and extension to degrade the race. Consequently, the mass of the population in our large cities will not bear a comparison with that of savage nations, in which if imperfect or deformed individuals should survive the hardships of their first rearing, they are prevented by the kind of aversion they inspire, from propagating their deformities. The Hottentots have become almost proverbial for ugliness; and one of their tribes, the Bosjesmen, are plainly ranked by an acute and intelligent traveller "among the ugliest of human beings[60]." The numerous sketches of Bosjesmen and Hottentots taken by Mr. S. Daniel, have been very kindly and politely shewn to me by his brother Mr. W. Daniel. In form, variety, and expression of countenance, they are not at all inferior to our cocknies; while in animation, in beauty, symmetry and strength of body, in ease and elegance of attitude, they are infinitely superior.

This inattention to breed is not, however, of so much consequence in the people as in the rulers; in those to whom the destinies of nations are intrusted; on whose qualities and actions depend the present and future happiness of millions. Here, unfortunately, the evil is at its height: laws, customs, prejudices, pride, bigotry, confine them to intermarriages with each other, and thus degradation of race is added to all the pernicious influences inseparable from such exalted stations. What result

[60] Barrow, *Travels in Southern Africa;* v. 1, p. 277.

should we expect if a breeder of horses or dogs were restricted in his choice to some ten or twenty families taken at random? if he could not step out of this little circle to select finely formed or high-spirited individuals? How long a time would elapse before the fatal effects of this in-breeding would be conspicuous in the degeneracy of the descendants? The strongest illustration of these principles will be found in the present state of many royal houses in Europe: the evil must be progressive, if the same course of proceeding be continued.

I shall cite a single example to prove, what will to most persons seem unnecessary, namely, that mental defects are propagated as well as corporeal. "We know," says HALLER, "a very remarkable instance of two noble females, who got husbands on account of their wealth, although they were nearly idiots, and from whom this mental defect has extended for a century into several families, so that some of all their descendants still continue idiots in the fourth and even in the fifth generation[61]."

[61] *Elem. Phisiol.* lib. 29, sect. ii. §. 8.

Part II

ARTIFICIAL SELECTION AND THE DEVELOPMENT OF DARWIN'S THEORY OF NATURAL SELECTION

Editor's Comments on Papers 9 Through 13

9 PRICHARD
Excerpt from *Book IX: General Survey of the Causes Which Have Produced Varieties in the Human Species, with Remarks on the Origin of Nations and on the Diversity of Languages*

10 LYELL
Excerpt from *Principles of Geology, Being an Attempt to Explain the Former Changes of the Earth's Surface, by Reference to Causes Now in Operation*

11 YOUATT
Excerpt from *Breeding.—Parturition.*

12 SEBRIGHT
The Art of Improving the Breeds of Domestic Animals. In a Letter to the Right Hon. Sir Joseph Banks, K.B.

13 WILKINSON
Excerpt from *Remarks on the Improvement of Cattle, &c. In a Letter to Sir John Saunders Sebright, Bart. M.D.*

The theory that adaptive evolution occurs as the consequence of natural, sexual, and artificial selection operating on heritable variations was advocated by Charles Darwin (1859) in *On the Origin of Species by Means of Natural Selection, or the Preservation of Favoured Races in the Struggle for Life*. Darwin had served as naturalist aboard the H.M.S. *Beagle* during its 1831–1836 world-encircling voyage. As Darwin began to prepare his journal of the voyage for publication and to study his specimen collections (especially those from the Galapagos Islands), he began to realize that numerous facts indicated the common descent of species. This led Darwin to open the first of a series of *Notebooks on Transmutation of Species* in July 1837 in which he recorded "any facts that might bear on the question" (Darwin, 1837, cited in Darwin and Seward, 1903, I:367).

Darwin's attention was probably drawn to the study of variation of plants and animals under domestication by the following passage that he read in 1837 and marked in his copy of the fifth edition of Charles Lyell's *Principles of Geology Being an Attempt to Explain Former Changes of the Earth's Surface by Reference to Causes Now in Operation:* "The best authenticated examples of the extent to which species can be made to vary, may be looked for in the history of domesticated animals and plants" (Lyell, 1837; see also Paper 10). Charles Lyell's attention was drawn to the role that selective breeding played in accumulating variations in domesticated animals and the conservative dimension of selection operating in nature when he read the second edition of James Prichard's *Researches into the Physical History of Mankind* (1826, see Paper 9). Charles Darwin had read the first edition of Lyell's *Principles of Geology* while on board the H.M.S. *Beagle*. Darwin was greatly impressed with Charles Lyell's success in applying the uniformity-of-causes principle (the theory that present causes now in operation are sufficient to explain past changes) to scientifically interpret the geological history of the earth. Darwin used this principle to help guide his search for the forces in nature that were responsible for bringing about the transmutation of species.*

Charles Darwin learned about artifical selection theory and the importance of selection in bringing about heritable changes in a population by reading the animal breeding literature (Darwin, 1868:10; Ruse, 1975). Darwin, writing in his autobiography, recalled that he began collecting

> facts on a wholesale scale, more especially with respect to Domesticated productions, by printed enquiries, by conversation with skillful breeders and gardeners, and by extensive reading . . . I soon perceived that selection was the keystone of man's success in making useful races of animals and plants. But how selection could be applied to organisms living in a state of nature remained for sometime a mystery to me. (Darwin in Barlow, 1958:119–120).

Charles Darwin also had the opportunity to observe personally the effects of selective breeding of domesticated animals during this period (1837–1838). His uncle Josiah Wedgewood was one of England's leading sheep breeders and his own Darwin family were pigeon fanciers (Ruse, 1979:178; Secord, 1981), a hobby that

*Darwin formulated a number of nonselectionist theories of evolution before constructing his theory of selection to explain the origin of species (Gruber and Barrett, 1974; Kohn, 1980).

Charles was to take up later in an attempt to better understand the origin of variation and its hereditary transmission.

The status of selective breeding theory during the 1830s can be ascertained, in part, by reading the views of William Youatt who wrote several books on animal husbandry. The principle of selection was defined by Youatt (1837:60) as "that which enables the argriculturalist not only to modify the character of his flock, but to change it altogether—the magician's wand, by means of which he may summon into life whatever form and mould he pleases. . . ." William Youatt and numerous other breeders emphasized that every individual was unique and that if a population was to be improved, extreme care had to be taken in selecting both the sires (males) and dams (females) from which to breed the next generation (Mayr, 1977:325).

Youatt's books on horses (1831), cattle (1834), sheep (1837), and pigs (1847) were important sources of information for Darwin on the breeding of farm animals (Wood, 1973). Youatt's 1834 views on the role that selection should play in cattle breeding are reprinted in Paper 11.

Darwin's reading list indicates that he did not read any of Youatt's books until 1840—two years after he had constructed his theory of evolution by natural selection (Vorzimmer, 1977). Darwin used information on animal breeding contained in Youatt's books in writing his *Sketch of 1842* (de Beer, 1958), *Essay of 1844* (de Beer, 1958), Linnean Society paper (1858), *On the Origin of Species* (1859) and *Variation of Animals and Plants Under Domestication* (1868). Darwin, in a November 27, 1859, letter to T. H. Huxley, wrote:

> About breeding, I know of no one book. I did not think well of Lowe [D. Low, *The Breeds of Domestic Animals of the British Isles* 1842]. But I can name none better. Youatt I look at as a far better and *more practical* authority; but then his views and facts are scattered through three or four thick volumes. I have picked up most by reading really numberless special treatises and *all* agricultural and horticultural journals; but it is the work of long years. *The difficulty is to know what to trust* (Darwin, 1899, II:75).

Two pamphlets on the selective breeding of animals that Charles Darwin read during the first part of 1838 are thought to have played a crucial role in making him aware of the fact that selective processes analogous to artificial selection were operating in the rest of nature (Ruse, 1975). John Sebright's 1809 pamphlet on *The Art of Improving the Breeds of Domestic Animals . . . (Paper* 12) contains not only a review of selective breeding in domestic animals but, more importantly, a brief discussion of selection operating in the rest of nature (pp. 15–16). Sebright gave examples of what Darwin was to later classify as natural selection and sexual selection. Darwin had the analogy between artifical selection and selection in the rest of nature thrust right at him when he read the Sebright pamphlet (Ruse, 1975). Charles Darwin's personal annotated copy of Sebright's pamphlet is reprinted as Paper 12.

Darwin wrote in his second *Notebook on the Transmutation of Species* that the Sebright pamphlet contained "excellent observations of sickly offspring being cut off so that not propagated by nature . . . and that whole art of making varieties may be inferred from the facts stated" (de Beer, 1960–1967, II, C:133; Ruse, 1975). In a footnote to the preceding entry, Darwin wrote that Sebright's views were "fully supported by Mr. Wilkinson."

John Wilkinson's 1820 pamphlet entitled *Remarks on the Improvement of Cattle, etc. in a Letter to Sir John Saunders Sebright, Bart, M.P.* (excerpted as Paper 12) contained the passage: "The longer also these perfections [of domesticated breeds] have been continued, the more stability will they have acquired, and the more will they partake of nature itself." Darwin put quotes around this sentence and drew two lines beside it in his copy of the pamphlet (Ruse, 1975). This hopeful comment by Wilkinson on the lasting effects of artificial selection contrasted with the widespread belief in the impermanence of the effects brought about by artificial selection held by numerous breeders including Sebright (Paper 12).

The anonymous (1838) review, attributed to David Brewster, of August Comte's *Cours de Philosophie Positive,* which Charles Darwin read between August 7 and 12, 1838, had a major impact on Darwin's thinking about natural and artificial selection (Schweber, 1977). Comte's views on experiment as a powerful instrument of investigation was summarized by Brewster (1838:303): "This instrument is experiment, by means of which we observe bodies out of their natural state; by placing them in artificial aspects and conditions contrived for exhibiting to us, under the most favourable circumstances, their phenomena and their properties." Brewster's review of Comte refocused Darwin's attention on "picking" by breeders and caused Darwin to consider the possibility that such artificial selection might play the role of "experiment" in unraveling the complex phenomena involved in the "transmutation of species" (Schweber, 1978:293).

Brewster in his review also discussed the importance that Comte attached to the numerical verification of hypotheses

(Brewster, 1838:299). This led Darwin to begin searching for a quantitative statement of some aspect of the theory of evolution he was trying to construct (Schweber, 1977:264). Darwin found such a quantitative statement when he read the sixth edition of *An Essay on the Principle of Population* by Robert Thomas Malthus (1826:6) in September 1838: "It may safely be pronounced, therefore, that the [human] population, when unchecked, goes on doubling itself every twenty-five years, or increases in a geometric ratio."

The thoughts on the analogy between artificial selection ("picking") and what was happening in nature that Darwin recorded in his *Notebooks on the Transmutation of Species* before and after he read the selective breeding pamphlets of Sebright and Wilkinson provide no evidence to support the contention that their ideas concerning selection had any immediate impact on Darwin's thinking. Darwin, in fact, appears to have denied the existence of such an analogy both before and after reading the selective breeding pamphlets as the following two most relevant entries in his *Notebooks* seem to indicate:

> The changes in species must be very slow owing to physical changes slow & offspring not picked—as men do when making varieties. (de Beer, 1960–1967, II, C:17)

> The varieties of domesticated animals must be most complicated, because they are partly local & then the local ones taken to fresh country & breed confined certain best individuals—scarcely any breed but what some individuals are picked out,—in a really natural breed, not one is picked out. . . . (de Beer, 1960–1967, III, D:20)

It is possible that Darwin made these entries not to deny the analogy but rather to help him remember arguments that could be used against the argument by analogy from artificial selection. Darwin, writing in his *Autobiography*, stated that he had made a habit of recording views opposed to his as he "had found by experience that such facts and thoughts were more apt to escape from the memory than favourable ones." (Darwin in Barlow, 1958:123). It is also possible that those pages that Darwin excised from his *Notebooks* for use in writing the *Origin* volume and which still have not been found contain Darwin's thoughts in support of arguing by analogy from "picking" to what was happening in the rest of nature.

While the writing of Sebright and others had made Darwin aware of the fact that selection was operating in nature, Charles Darwin did not consider the natural means of selection to be powerful enough to bring about much if any evolution until he read Malthus' *Essay*. Charles Darwin has summarized the role that Malthusian population theory played in the historical development of his theory of evolution by natural selection:

> When I visited, during the voyage of H.M.S. *Beagle*, the Galapagos Archipelago, situated in the Pacific Ocean about 500 miles from the shore of South America, I found myself surrounded by peculiar species of birds, reptiles, and plants, existing nowhere else in the world. Yet they nearly all bore an American stamp. . . . I often asked myself how these many peculiar animals and plants had been produced: the simplest answer seemed to be that the inhabitants of the several islands had descended from each other, undergoing modification in the course of their descent; and that all the inhabitants of the archipelago had descended from those of the nearest land, namely America, whence colonists would naturally have been derived. But it long remained to me an inexplicable problem how the necessary degree of modification could have been effected, and it would have thus remained for ever, had I not studied domestic productions, and thus acquired a just idea of the power of Selection. As soon as I had fully realized this idea, I saw, on reading Malthus on Population, that Natural Selection was the inevitable result of the rapid increase of all organic beings; for I was prepared to appreciate the struggle for existence by having long studied the habits of animals. (Darwin, 1868:9–10)

Malthus, by stating that a population tends to increase in a geometric ratio, enabled Darwin to perceive just how powerful a force selection operating in nature really was. The power of selection was generated by the overproduction of offspring, which greatly intensified the "struggle for existence" in nature. The contribution that Malthus made to the development of Darwin's ideas concerning the importance of natural selection as an ecological force bringing about "descent with modification" has been succinctly summarized by Limoges (1970:79):

> What Malthus provided Darwin was not the idea of the struggle for existence, which was by then common. But rather the notion of the intensity of that struggle, of its constraining power on living beings, the idea that a geometric increase implies that a constant pressure is exerted on living beings which necessarily engenders among them an incessant war—the ancestral form of what is called "population pressure" in present day genetics. (English translation in Schweber 1978: 324)

The first record of Darwin's positive views on the analogy between the way that varieties of domesticated species are made by man and the way that species are produced in nature is contained in the following entry Darwin made in his *Fourth Notebook on the Transmutation of Species* during December 1838: "It is a beautiful part of my theory, that domesticated races of organics are made by precisely same means as species—but latter far more perfectly and infinitely slower" (de Beer, 1960–1967, IV, E:71). Charles Darwin was so anxious to avoid having his newly constructed theory of selection prejudice his thoughts that he "determined not for some time to write even the briefest sketch of it." (Darwin in Barlow, 1958:120)

In 1839 Darwin composed an eight-page questionnaire entitled *Questions About the Breeding of Animals* that he had printed and sent to animal breeders and naturalists (Darwin, 1839; Freeman and Gautrey, 1969; Vorzimmer, 1969). Darwin continued to use the word "picking" (questions 3 and 6) to describe what he was later to call "artificial selection."

Darwin finally wrote a brief 35-page *Sketch* of his theory in 1842. He used the term "selection" for the first time in this *Sketch* to describe what he had been calling "picking," "destruction," "preservation," and "sift." Two years later, he expanded the *Sketch* into his 230-page *Essay of 1844*. Darwin, in Part I of his *Essay*, argued from the analogy of the role of man's selection in the production of varieties under domestication to the circumstances leading to the production of wild varieties to establish the possibility of his theory (Manier, 1978:7).

Darwin drew an analogy between artificial selection and both of the forms of selection he perceived to be operating in the rest of nature—selection by death (natural selection) and selection in time of fullest vigor, which involved struggle for mates (sexual selection), in both the *Sketch of 1842* and the *Essay of 1844*.

Charles Darwin was able to use his knowledge about variation in domesticated species to construct his theory of the natural means of selection because he was already convinced that the transmutation of species was a fact before he began reading the breeding literature. Most naturalists, however, were not even thinking about the processes by which transmutation of species might have occurred.

The study of domesticated varieties provided naturalists with one of the strongest arguments against the idea that transmutation of species had occurred. Domestic varieties, when turned wild, were observed by naturalists to either go extinct or revert back to the original type (see, for example, Lyell, 1832, II:28, 32; see also Paper 10). The English naturalist Edward Blyth (1835:46), who perceived how selection could maintain (but not change) the qualities of a species in nature wrote:

> The same law, therefore, which was intended by Providence to keep up the typical qualities of a species, can be easily converted by man into a means of raising different varieties; but it is also clear that, if man did not keep up these breeds by regulating the sexual intercourse, they would all naturally soon revert to the original type.

Alfred Russel Wallace, the cofounder of the Darwin-Wallace theory of adaptive evolution by selection, considered this argument in favor of the fixity of species to be so important that he devoted a significant part of his 1858 essay "On the Tendency of Varieties to Depart Indefinitely from the Original Type" to summarizing and then refuting the argument (Wallace, 1858:53–54, 59–61). Wallace contended that the same principle (selection) that generates change in species "in a state of nature will also explain why domestic varieties have a tendency to revert to the original type."

Charles Darwin did not publicly advocate his theory of adaptive evolution by selection until 1858. Alfred R. Wallace sent his 1858 manuscript, "On the Tendency of Varieties to Depart Indefinitely from the Original Type," to Charles Darwin. This triggered a chain of events that led to the publication of a five-paragraph extract of Darwin's manuscript on species along with Wallace's manuscript in the August 1858 issue of the *Journal of the Proceedings of the Linnean Society, Zoology.*

Alfred Wallace constructed the theory of adaptive evolution by selection independently of Charles Darwin (McKinney, 1972). Wallace constructed his theory in February 1858 while thinking about a possible mode for the origin of new species. He wrote the following summary of the chain of his thoughts in constructing the theory forty-five years afterward:

> Somehow my thoughts turned to the "positive checks" [disease, famine, accidents, war, etc.] to increase among savages and others described . . . in the celebrated *Essay on Population*, by Malthus. . . . It suddenly occurred to me that in the case of wild animals these checks would act with much severity, and . . . that those individuals which every year were removed by these causes—termed collectively the "struggle for existence"—must on the average and in the long run be inferior in some one or more ways to those which managed to survive. (Wallace, 1903:78)

Darwin, amazed at the similarity between Wallace's manuscript and his own summary of the adaptive evolution by selection theory, wrote Charles Lyell that: "I never saw a more striking coincidence; if Wallace had my MS. sketch written out in 1842, he could not have made a better short abstract! Even his terms now stand as heads of my chapters" (Darwin, June 18, 1858 letter to Charles Lyell in F. Darwin, 1899, I:473). Darwin wrote Lyell again one week later stating: "There is nothing in Wallace's sketch which is not written out much fuller in my sketch, copied out in 1844, and read by Hooker some dozen years ago" (Darwin, June 25, 1858, letter to Charles Lyell in F. Darwin, 1899, I:474). In the same letter Darwin wrote that Wallace and he differed ". . . only, [in] that I was led to my views from what artificial selection has done for domestic animals."

9

Reprinted from pages 557–558 of *Researches into the Physical History of Mankind,* 2d ed., vol. II, London, 1826

BOOK IX: GENERAL SURVEY OF THE CAUSES WHICH HAVE PRODUCED VARIETIES IN THE HUMAN SPECIES, WITH REMARKS ON THE ORIGIN OF NATIONS AND ON THE DIVERSITY OF LANGUAGES

J. C. Prichard

[*Editor's Note:* In the original, material procedes and follows this excerpt.]

It is generally supposed that cultivation is the most productive cause of varieties in the kind, both in the animal and vegetable kingdom. But it may be questioned, does cultivation actually give rise to entirely new varieties, or does it only foster and propagate those which have sprung up naturally, or as it is termed accidentally?

In this latter way the influence of art is very important in constituting breeds, as of cattle, dogs, horses. The artificial process consists in a careful selection of those individual animals which happen to be possessed, in a greater degree than the generality, of any particular characters which it is desirable to perpetuate. These are kept for the propagation of the stock, and a repeated attention is paid to the same circumstances, till, the effect continually increasing, a particular figure, colour, proportion of limbs, or any other attainable quality, is established in the race, and the uniformity of the breed is afterwards maintained by removing from it any new variety which may casually spring up in it.

But whether animals in a domesticated state, independently of this sort of control, are more disposed to exhibit new varieties in their offspring, or to bring forth a progeny of different form or colour from the breed, is a question which I know not how to determine. Animals however, in the state of domestication, are placed under circumstances so widely different from those of their natural and ordinary state, that this altered condition seems very likely to occasion deviations in their progeny.

But there is no other cause which, in so important a manner, alters the circumstances under which the fœtus is bred, as a change of climate, and accordingly it seems undoubted, that changes of climate have given rise to deviations in the breed of animals. It is impossible, otherwise, to account for certain peculiarities in various races, which exist only in particular places, where they appear to have originated, and to which they are in a great measure confined. As this consideration is a very important one with respect to the theory of natural varieties, I shall mention, before I proceed further, several instances, in which it appears that a deviation from the general character of a race, either of men or inferior animals, has taken place in connexion with external circumstances, chiefly local.

10

Reprinted from pages 26–35 of *Principles of Geology, Being an Attempt to Explain the Former Changes of the Earth's Surface, by Reference to Causes Now in Operation*, vol. II, John Murray, London, 1832

PRINCIPLES OF GEOLOGY

C. Lyell

[*Editor's Note:* In the original, material precedes and follows this excerpt.]

Now let us first inquire what positive facts can be adduced in the history, of known species, to establish a great and permanent amount of change in the form, structure, or instinct of individuals descending from some common stock. The best authenticated examples of the extent to which species can be made to vary, may be looked for in the history of domesticated animals and cultivated plants. It usually happens that those species, both of the animal and vegetable kingdom, which have the greatest pliability of organization, those which are most capable of accommodating themselves to a great variety of new circumstances, are most serviceable to man. These only can be carried by him into different climates, and can have their properties or instincts variously diversified by differences of nourishment and habits. If the resources of a species be so limited, and its habits and faculties be of such a confined and local character, that it can only flourish in a few particular spots, it can rarely be of great utility.

We may consider, therefore, that in perfecting the arts of domesticating animals and cultivating plants, mankind have first selected those species which have the most flexible frames and constitutions, and have then been engaged for ages in conducting a series of experiments, with much patience and at great cost, to ascertain what may be the greatest possible deviation from a common type which can be elicited in these extreme cases.

The modifications produced in the different races of dogs, exhibit the influence of man in the most striking point of view. These animals have been transported into every climate, and placed in every variety of circumstances; they have been made, as a modern naturalist observes, the servant, the companion, the guardian, and the intimate friend of man, and the power of a superior genius has had a wonderful influence, not only on their forms, but on their manners and intelligence*. Different races have undergone remarkable changes in the quantity and colour of their clothing: the dogs of Guinea are almost naked, while those of the Arctic circle are covered with a warm coat both of hair and wool, which enables them to bear the most intense cold without inconvenience. There are differences also of another kind no less remarkable, as in size, the length of their muzzles, and the convexity of their foreheads.

But if we look for some of those essential changes which would be required to lend even the semblance of a foundation for the theory of Lamarck, respecting the growth of new organs and the gradual obliteration of others, we find nothing of the kind. For in all these varieties of the dog, says Cuvier, the relation of the bones with each other remain essentially the same; the form of the teeth never changes in any perceptible degree, except that in some individuals, one additional false grinder occasionally appears, sometimes on the one side, and sometimes on the other†. The greatest departure from a common type, and it constitutes the maximum of variation as yet known in the animal kingdom, is exemplified in those races of dogs which have a supernumerary toe on the hind foot with the corresponding tarsal bones, a variety analogous to one presented by six-fingered families of the human race‡.

Lamarck has thrown out as a conjecture, that the wolf may have been the original of the dog, but he has adduced no data to bear out such an hypothesis. "The wolf," observes Dr. Prichard, "and the dog differ, not only with respect to their habits and instincts, which in the brute creation are very uniform within the limits of one species; but some differences have also been pointed out in their internal organization,

* Dureau de la Malle, Ann. des. Sci. Nat. tom. xxi. p. 63. Sept. 1830.

† Disc. Prel., p. 129, sixth edition. ‡ Ibid.

particularly in the structure of a part of the intestinal canal*."

It is well known that the horse, the ox, the boar and other domestic animals, which have been introduced into South America, and have run wild in many parts, have entirely lost all marks of domesticity, and have reverted to the original characters of their species. But the dog has also become wild in Cuba, Hayti, and in all the Caribbean islands. In the course of the seventeenth century, they hunted in packs from twelve to fifty, or more in number, and fearlessly attacked herds of wild-boars and other animals. It is natural, therefore, to enquire to what form they reverted? Now they are said by many travellers to have resembled very nearly the shepherd's dog; but it is certain that they were never turned into wolves. They were extremely savage, and their ravages appear to have been as much dreaded as those of wolves, but when any of their whelps were caught, and brought from the woods to the towns, they grew up in the most perfect submission to man.

As the advocates of the theory of transmutation trust much to the slow and insensible changes which time may work, they are accustomed to lament the absence of accurate descriptions, and figures of particular animals and plants, handed down from the earliest periods of history, such as might have afforded data for comparing the condition of species, at two periods considerably remote. But fortunately, we are in some measure independent of such evidence, for by a singular accident, the priests of Egypt have bequeathed to us, in their cemeteries, that information, which the museums and works of the Greek philosophers have failed to transmit.

For the careful investigation of these documents, we are greatly indebted to the skill and diligence of those naturalists who accompanied the French armies during their brief occupation of Egypt: that conquest of four years, from which we may date the improvement of the modern Egyptians in the arts

* Prichard, Phys. Hist. of Mankind, vol i. p. 96, who cites Professor Güldenstädt.

and sciences, and the rapid progress which has been made of late in our knowledge of the arts and sciences of their remote predecessors. Instead of wasting their whole time as so many preceding travellers had done, in exclusively collecting human mummies, M. Geoffroy and his associates examined diligently, and sent home great numbers of embalmed bodies of consecrated animals, such as the bull, the dog, the cat, the ape, the ichneumon, the crocodile, and the ibis.

To those who have never been accustomed to connect the facts of Natural History with philosophical speculations, who have never raised their conceptions of the end and import of such studies beyond the mere admiration of isolated and beautiful objects, or the exertion of skill in detecting specific differences, it will seem incredible that amidst the din of arms, and the stirring excitement of political movements, so much enthusiasm could have been felt in regard to these precious remains.

In the official report drawn up by the Professors of the Museum at Paris, on the value of these objects, there are some eloquent passages which may appear extravagant, unless we reflect how fully these naturalists could appreciate the bearing of the facts thus brought to light on the past history of the globe.

"It seems," say they, "as if the superstition of the ancient Egyptians had been inspired by Nature, with a view of transmitting to after ages a monument of her history. That extraordinary and whimsical people, by embalming with so much care the brutes which were the objects of their stupid adoration, have left us, in their sacred grottoes, cabinets of zoology almost complete. The climate has conspired with the art of embalming to preserve the bodies from corruption, and we can now assure ourselves by our own eyes what was the state of a great number of species three thousand years ago. We can scarcely restrain the transports of our imagination, on beholding thus preserved with their minutest bones, with the smallest portions of their skin, and in every particular most perfectly recognizable, many an animal, which at Thebes or Memphis, two or three thousand years ago, had its own priests and altars *."

Among the Egyptian mummies thus procured were not only those of numerous wild quadrupeds, birds, and reptiles, but, what was perhaps of still greater importance in deciding the great question under discussion, there were the mummies of domestic animals, among which those above mentioned, the bull, the dog, and the cat, were frequent. Now such was the conformity of the whole of these species to those now living, that there was no more difference, says Cuvier, between them than between the human mummies and the embalmed bodies of men of the present day. Yet some of these animals have since that period been transported by man to almost every variety of climate, and forced to accommodate their habits to new circumstances, as far as their nature would permit. The cat, for example, has been carried over the whole earth, and, within the last three centuries, has been naturalized in every part of the new world, from the cold regions of Canada to the tropical plains of Guiana; yet it has scarcely undergone any perceptible mutation, and is still the same animal which was held sacred by the Egyptians.

Of the ox, undoubtedly there are many very distinct races; but the bull Apis, which was led in solemn processions by the Egyptian priests, did not differ from some of those now living. The black cattle that have run wild in America, where there were many peculiarities in the climate not to be found, perhaps, in any part of the old world, and where scarcely a single plant on which they fed was of precisely the same species, instead of altering their form and habits, have actually reverted to the exact likeness of the aboriginal wild cattle of Europe.

In answer to the arguments drawn from the Egyptian mummies, Lamarck said that they were identical with their living descendants in the same country, because the climate and

* Ann. du Museum, d'Hist. Nat., tom. i. p. 234. 1802. The reporters were MM. Cuvier, Lacépède, and Lamarck.

physical geography of the banks of the Nile have remained unaltered for the last thirty centuries. But why, we ask, have other individuals of these species retained the same characters in so many different quarters of the globe, where the climate and many other conditions are so varied?

The evidence derived from the Egyptian monuments was not confined to the animal kingdom; the fruits, seeds, and other portions of twenty different plants, were faithfully preserved in the same manner; and among these the common wheat was procured by Delille, from closed vessels in the sepulchres of the kings, the grains of which retained not only their form, but even their colour, so effectual has proved the process of embalming with bitumen in a dry and equable climate. No difference could be detected between this wheat and that which now grows in the East and elsewhere, and similar identifications were made in regard to all the other plants.

And here we may observe, that there is an obvious answer to Lamarck's objection *, that the botanist cannot point out a country where the common wheat grows wild, unless in places where it may have been derived from neighbouring cultivation. All naturalists are well aware that the geographical distribution of a great number of species is extremely limited, and that it was to be expected that every useful plant should first be cultivated successfully in the country where it was indigenous, and that, probably, every station which it partially occupied, when growing wild, would be selected by the agriculturist as best suited to it when artificially increased. Palestine has been conjectured, by a late writer on the Cerealia, to have been the original habitation of wheat and barley, a supposition which appears confirmed by Hebrew and Egyptian traditions, and by tracing the migrations of the worship of Ceres, as indicative of the migrations of the plant †.

If we are to infer that some one of the wild grasses has been

* Phil. Zool., tom. i., p. 227.

† L'Origine et la Patrie des Céréales, &c. Ann. des Sci. Nat., tom. ix., p. 61.

transformed into the common wheat, and that some animal of the genus *canis*, still unreclaimed, has been metamorphosed into the dog, merely because we cannot find the domestic dog, or the cultivated wheat, in a state of nature, we may be next called upon to make similar admissions in regard to the camel; for it seems very doubtful whether any race of this species of quadruped is now wild.

But if agriculture, it will be said, does not supply examples of extraordinary changes of form and organization, the horticulturist can, at least, appeal to facts which may confound the preceding train of reasoning. The crab has been transformed into the apple; the sloe into the plum: flowers have changed their colour and become double; and these new characters can be perpetuated by seed,—a bitter plant with wavy sea-green leaves has been taken from the sea-side where it grew like wild charlock, has been transplanted into the garden, lost its saltness, and has been metamorphosed into two distinct vegetables as unlike each other as is each to the parent plant—the red cabbage and the cauliflower. These, and a multitude of analogous facts, are undoubtedly among the wonders of nature, and attest more strongly, perhaps, the extent to which species may be modified, than any examples derived from the animal kingdom. But in these cases we find, that we soon reach certain limits, beyond which we are unable to cause the individuals, descending from the same stock, to vary; while, on the other hand, it is easy to show that these extraordinary varieties could seldom arise, and could never be perpetuated in a wild state for many generations, under any imaginable combination of accidents. They may be regarded as extreme cases brought about by human interference, and not as phenomena which indicate a capability of indefinite modification in the natural world.

The propagation of a plant by buds or grafts, and by cuttings, is obviously a mode which nature does not employ; and this multiplication, as well as that produced by roots and layers, seems merely to operate as an extension of the life of an individual, and not as a reproduction of the species, as happens by seed. All plants increased by the former means retain precisely the peculiar qualities of the individual to which they owe their origin, and, like an individual, they have only a determinate existence; in some cases longer and in others shorter *. It seems now admitted by horticulturists, that none of our garden varieties of fruit are entitled to be considered strictly permanent, but that they wear out after a time †; and we are thus compelled to resort again to seeds; in which case, there is so decided a tendency in the seedlings to revert to the original type, that our utmost skill is sometimes baffled in attempting to recover the desired variety.

The different races of cabbages afford, as we have admitted, an astonishing example of deviation from a common type; but we can scarcely conceive them to have originated, much less to have lasted for several generations, without the intervention of man. It is only by strong manures that these varieties have been obtained, and in poorer soils they instantly degenerate. If, therefore, we suppose in a state of nature the seed of the wild Brassica oleracea to have been wafted from the sea-side to some spot enriched by the dung of animals, and to have there become a cauliflower, it would soon diffuse its seed to some comparatively steril soils around, and the offspring would relapse to the likeness of the parent stock, like some individuals which may now be seen growing on the cornice of old London bridge.

But if we go so far as to imagine the soil, in the spot first occupied, to be constantly manured by herds of wild animals, so as to continue as rich as that of a garden, still the variety could not be maintained, because we know that each of these races is prone to fecundate others, and gardeners are compelled to exert the utmost diligence to prevent cross-breeds. The intermixture of the pollen of varieties growing in the poorer soil around, would soon destroy the peculiar characters

* Smith's Introduction to Botany, p. 138. Edit. 1807.

† See Mr. Knight's Observations, Hort. Trans., vol. ii., p. 160.

of the race which occupied the highly-manured tract; for, if these accidents so continually happen in spite of us, among the culinary varieties, it is easy to see how soon this cause might obliterate every marked singularity in a wild state.

Besides, it is well-known that although the pampered races which we rear in our gardens for use or ornament, may often be perpetuated by seed, yet they rarely produce seed in such abundance, or so prolific in quality, as wild individuals; so that, if the care of man were withdrawn, the most fertile variety would always, in the end, prevail over the more steril.

Similar remarks may be applied to the double flowers which present such strange anomalies to the botanist. The ovarium, in such cases, is frequently abortive, and the seeds, when prolific, are generally much fewer than where the flowers are single.

Some curious experiments recently made on the production of blue instead of red flowers in the Hydrangea hortensis, illustrate the immediate effect of certain soils on the colours of the petals. In garden-mould or compost, the flowers are invariably red; in some kinds of bog-earth they are blue; and the same change is always produced by a particular sort of yellow loam.

Linnæus was of opinion that the primrose, oxlip, cowslip, and polyanthus, were only varieties of the same species. The majority of modern botanists, on the contrary, consider them to be distinct, although some conceived that the oxlip might be a cross between the cowslip and the primrose. Mr. Herbert has lately recorded the following experiment:—" I raised from the natural seed of one umbel of a highly-manured red cowslip, a primrose, a cowslip, oxlips of the usual and other colours, a black polyanthus, a hose-in-hose cowslip, and a natural primrose bearing its flower on a polyanthus stalk. From the seed of that very hose-in-hose cowslip I have since raised a hose-in-hose primrose. I therefore consider all these to be only local varieties depending upon soil and situation *." Pro-

* Hort. Trans., vol. iv., p. 19.

fessor Henslow, of Cambridge, has since confirmed this experiment of Mr. Herbert, so that we have an example, not only of the remarkable varieties which the florist can obtain from a common stock, but of the distinctness of analogous races found in a wild state *.

On what particular ingredient, or quality in the earth, these changes depend, has not yet been ascertained †. But gardeners are well aware that particular plants, when placed under the influence of certain circumstances, are changed in various ways according to the species; and as often as the experiments are repeated similar results are obtained. The nature of these results, however, depends upon the species, and they are, therefore, part of the specific character; they exhibit the same phenomena again and again, and indicate certain fixed and invariable relations between the physiological peculiarities of the plant, and the influence of certain external agents. They afford no ground for questioning the instability of species, but rather the contrary; they present us with a class of phenomena which, when they are more thoroughly understood, may afford some of the best tests for identifying species, and proving that the attributes originally conferred, endure so long as any issue of the original stock remains upon the earth.

* Loudon's Mag. of Nat. Hist., Sept. 1830, vol. iii., p. 408.
† Hort. Trans., vol. iii., p. 173.

11

Reprinted from pages 522–526 and 527 of *Cattle; Their Breeds, Management, and Diseases*, Baldwin and Cradock, London, 1834

BREEDING.—PARTURITION.

W. Youatt

THE characteristics of the different breeds of British cattle, the peculiar excellencies and the peculiar defects of each, and their comparative value, as adapted to different climates and soil and pasture, have been already considered: a few remarks on *the principles* of breeding were reserved for this chapter.

That which lies at the foundation of the improvement of every stock, or the successful management of it, is the fact,—the common, but too much neglected axiom, that "*like produces like.*" This is the governing law in every portion of animated nature. There is not a deviation from it in the vegetable world, and the exceptions are few and far between among the lower classes of animals. When in the higher species the principle may not seem at all times to hold good, it is because another power, the intellectual—the imaginative—somewhat controls the mere organic one; or, in a great many instances, the organic principle is still in full activity, for the lost resemblance to generations gone by is pleasingly and strongly revived. The principle that "like produces like,*" was that which gave birth to the valuable, but too short-lived, new Leicester breed; it was the principle to which England is indebted for the short-horns, that are

* "The simple observation, that domestic animals possess a tendency to produce animals of a quality similar to their own, was the ground-work of all Bakewell's proceedings. It was equally obvious to others as to him, but by him first applied to the useful purpose to which it has since been rendered subservient. Having made this observation, he inferred, that by bringing together males and females possessing the same valuable properties, he should insure their presence in their offspring, probably in an increased degree, they being inherited from both parents; and he concluded, that by persisting in breeding from animals the produce of such selections, always keeping in sight the properties that constituted their value, he should at length establish a breed of cattle of which those properties would form the distinguishing and necessary characteristic. By this process it was that in his time, with respect to his long-horns, and subsequently with regard to other breeds of cattle, the term *blood* came to be distinctively applied. When reference could be made to a number of ancestors of distinguished excellence, the term *blood* was admitted."—The Rev. H. Berry's admirable Prize Essay on Breeding.

now establishing their superiority in every district of the kingdom. Every cow and heifer of the SHAKSPEARE blood could be recognized at first sight as having descended from Mr. Fowler's stock; and the admirer of the short-horns can trace in the best cattle of the present day the undoubted lineaments of FAVOURITE.

This principle extends to form, constitution, qualities, predisposition to, and exemption from disease, and to every thing that can render an animal valuable or worthless. It equally applies to the dam and to the sire. It is the foundation of scientific and successful breeding*.

Let it be supposed, that the cattle of a certain farmer have some excellent qualities about them; but there is a defect which considerably deteriorates from their value, and which he is anxious to remove. He remembers that "like produces like," and he looks about for a bull that possesses the excellence which he wishes to engraft on his own breed. He tries the experiment, and, to his astonishment, it is a perfect failure. His stock, so far from improving, have deteriorated.

The cause of this every-day occurrence was, that he did not fairly estimate the extent of the principle from which he expected so much. This new bull had the good point that was wanting in his old stock; but he too was deficient somewhere else, and, therefore, although his cattle had in some degree improved by him in one way, that was more than counterbalanced by the inheritance of his defects. Here is the secret of every failure—the grand principle of breeding. The new-comer, while he possesses that which was a desideratum in the old stock, should likewise possess every good quality that they had previously exhibited—then, and then alone, will there be improvement without alloy. What can a farmer expect if he sends a worthless cow to the best-bred bull—or, on the other hand, if his cows, although they may have many good qualities, are served by a bull that perhaps he has scarcely seen, or whose points he has not studied, and whose only recommendations are, that he is close at hand and may be had for little money?

The question as to the comparative influence of the sire and the dam is a difficult one to decide. That farmer will not err, who applies the grand principle of breeding equally to both of them. In the present system of breeding, most importance, and that very justly, is attributed to the male. He is the more valuable animal, and principally more valuable on account of the more numerous progeny that is to proceed from him, and thus his greater general influence; and therefore superior care is bestowed on the first selection of him for rearing. The farmer studies the bull-calf closely, and assures himself that he possesses, in a more than usual degree, the characteristic excellencies of the breed. When this care as to the possession of such combination of good points has extended from the sire to the son through several successive generations, it may be readily supposed that he will possess them in a higher degree than the female can. They

* There are a few strange exceptions to this, showing the power of imagination even over so dull a beast as the cow. Her progeny is often much affected by circumstances that happen during the time of conception, or rather during the period she is in season. Mr. Boswell says, "One of the most intelligent breeders I ever met with in Scotland, Mr. Mustard, of Angus, told me a singular fact with regard to what I have now stated. One of his cows chanced to come in season, while pasturing on a field, which was bounded by that of one of his neighbours, out of which an ox jumped, and went with the cow, until she was brought home to the bull. The ox was white, with black spots, and horned. Mr. Mustard had not a horned beast in his possession, nor one with any white on it. Nevertheless, the produce of the following spring was a black and white calf with horns.—Quarterly Journal of Agriculture, vol. i. Essays, p. 28.

will be made. as it were, a part and portion of his constitution, and he will acquire the power of more certainly, and to a greater extent, communicating them to his offspring.

In this way the influence of the sire may, in well-bred animals, be considered as superior to that of the female; but hers is always great, and must not be forgotten. In Arabia, where the mare is the object of chief attention, and her good qualities are carefully studied and systematically bred in her, the influence of the female decidedly preponderates; and, on the same principle, that of the highly bred cow will preponderate over that of the half-bred bull. Her excellencies are an hereditary and essential part of her, and more likely to be communicated to her offspring than those which have been only lately and accidentally acquired by the bull with no pedigree, or with many a blot in it. Custom and convenience, however, induce the generality of breeders to look most to the male. *

At the outset of his career, the farmer should have a clear and determined conception of the object that he wishes to accomplish. He should consider the nature of his farm; its abundance or deficiency of pasturage; the character of the soil; the seasons of the year when he will have plenty or deficiency of food; the locality of his farm; the market to which he has access, and the produce which will there be disposed of with greatest profit, and these things will at once point to him the kind of beast which he should be solicitous to obtain. The man of wealth and patriotism may have more extensive views, and nobly look to the general improvement of British cattle; but the farmer, with his limited means, and with the claims that press upon him, regards his cattle as a valuable portion of his own little property, and on which every thing should appear to be in natural keeping, and be turned to the best advantage.

The best beast for him is that which suits his farm the best; and, with a view to this, he studies, or ought to study, the points and qualities of his own cattle, and those of his neighbours. The dairy-man will regard the quantity of milk—the quality—the time that the cow continues in milk—its value for the production of butter or cheese—the character of the breed for quietness—or as being good nurses—the predisposition to red-water, garget, or dropping after calving—the natural tendency to turn every thing to nutriment—the easiness with which she is fattened, when given up as a milker, and the proportion of food requisite to keep her in full milk, or to fatten her when dry. The grazier will consider the kind of beast which his land will bear—the kind of meat most in demand in his neighbourhood—the early maturity—the quickness of fattening at any age—the quality of the meat—the parts on which the flesh and fat are principally laid—and, more than all, the hardihood and the adaptation of constitution to the climate and soil.

In order to obtain these valuable properties, the farmer will make himself perfectly master of the character and qualities of his own stock. He will trace the connexion of certain good qualities and certain bad ones, with an almost invariable peculiarity of shape and structure; and at length he will arrive at a clear conception, not so much of beauty of form (al-

* Mr. Adam Ferguson, of Woodhill, to whom the Highland Society of Scotland, and the Scottish agriculturists generally, are so much indebted, has an amusing anecdote on this point. "I recollect, several years ago, at a distinguished breeder's in Northumberland, meeting with a shrewd Scottish borderer, (indeed, if report be true, the original and identical Dinmont,) who, after admiring with a considerable spice of national pique, a very short-horned bull, demanded anxiously to see the dam. The cow being accordingly produced, and, having undergone a regular survey, Dandy vociferated, with characteristic *pith*, "I think naething of your bull now, wi' sic a caumb."—Quarterly Journal of Agriculture, vol. i. p. 34.

though that is a pleasing object to contemplate) as of that outline and proportion of parts with which *utility* is oftenest combined. Then carefully viewing his stock, he will consider where they approach to, and how far they wander from, this utility of form; and he will be anxious to preserve or to increase the one, and to supply the deficiency of the other.* He will endeavour to select from his own stock those animals that excel in the most valuable points, and particularly those which possess the greatest number of these points; and he will unhesitatingly condemn every beast that betrays manifest deficiency in any one important point. He will not, however, too long confine himself to his own stock, unless it is a very numerous one. The breeding from close affinities—the breeding *in and in*—has many advantages to a certain extent. It may be pursued until the excellent form and quality of the breed is developed and established. It was the source whence sprung the cattle and the sheep of Bakewell, and the superior cattle of Colling; and to it must also be traced the speedy degeneracy—the absolute disappearance of the new Leicester cattle, and, in the hands of many an agriculturist, the impairment of constitution and decreased value of the new Leicester sheep and the short-horned beasts. It has, therefore, become a kind of principle with the agriculturist to effect some change in his stock every second or third year, and that change is most conveniently effected by introducing a new bull. This bull should be, as nearly as possible, of the same sort; coming from a similar pasturage and climate; but possessing no relationship—or, at most, a very distant one—to the stock to which he is introduced. He should bring with him every good point which the breeder has laboured hard to produce in his stock, and, if possible, some improvement, and especially where the old stock may have been somewhat deficient; and most certainly he should have no manifest defect of form; and that most essential of all qualifications, a hardy constiution, should not be wanting.

There is one circumstance, however, which the breeder occasionally forgets, but which is of as much importance to the permanent value of his stock as any careful selection of animals can be—and that is, good keep. It was judiciously remarked by the author of the "Agricultural Report of Staffordshire," that "all good stock must be both bred with attention and well fed. It is necessary that these two essentials in this species of improvement should always accompany each other; for without good resources of keeping, it would be vain to attempt supporting a capital stock." This is true with regard to the original stock; it is yet more evident when animals are absurdly brought from a better to a poorer soil. The original stock

* "Upon the principle that 'like produces like,' he (Bakewell) started, and the advantages which crowned his exertions may be thus stated: an increased perfection of general symmetry, by which is to be understood not only a form attractive to the eye of taste, but one in which the judgment acknowledged a considerable preponderance of the valuable parts of the carcase over those of less value; an increased tendency to lay on flesh of a superior quality under all circumstances of feeding, and, of course, a superior article for the use of the consumer, produced by a decreased consumption of vegetable or other food.

"A person would often be puzzled: he would find different individuals possessing different perfections in different degrees—one, good flesh, and a tendency to fatten, with a bad form—another, with fine form, but bad flesh, and little disposition to acquire fat:—what rule should he lay down, by the observance of which good might be generally produced, and as little evil as possible effected?—UTILITY. The truly good form is that which secures constitution, health, and vigour—a disposition to lay on flesh, and with the greatest possible reduction of offal. Having obtained this, other things are of minor, although perhaps of considerable, importance."—The Rev. H. Berry's Prize Essay.

will deteriorate if neglected and half-starved; and the improved breed will lose ground even more rapidly, and to a far greater extent.

The full consideration, however, of the subject of breeding belongs to the work on "British Husbandry," and there full justice will be done to it: but the few hints that have here been dropped with reference to the fundamental principles on which the improvement of cattle must be founded will not, perhaps, be deemed irrelevant.*

THE PROPER AGE FOR BREEDING.

The proper age at which the process of breeding may be commenced will depend on various circumstances. Even with the early maturity of the short-horns, if the heifers could be suffered to run until they were two and a half, or three years old, they would become larger, finer, and more valuable; and their progeny would be larger and stronger: but the expense of the keep for so long a time is a question that must be taken into serious consideration. The custom which at one period was beginning to be so prevalent in the breeding districts, of putting the heifer to the male at one year old, or even at an earlier period, cannot be too much reprobated. At the time when they are most rapidly growing themselves, a sufficient quantity of nutriment cannot be devoted to the full development of the fœtus, and both the mother and the calf must inevitably suffer.

From two, to two-and-a-half years old, according to the quality of the pasture, will be the most advantageous time for putting the heifer to the bull. In fair pasture, the heifer will probably have attained sufficient growth at two years. If the period is prolonged after three years, and especially with good keep, the animal will often be in too high condition, and there will be much uncertainty as to her becoming pregnant.† At an

* The following extract from "the Rev. H. Berry's Prize Essay" contains the sum and substance of the principles of breeding:—

"A person selecting a stock from which to breed, notwithstanding he has set up for himself a standard of perfection, will obtain them with qualifications of different descriptions, and in different degrees. In breeding from such he will exercise his judgment, and decide what are indispensable or desirable qualities, and will cross with animals with a view to establish them. His proceeding will be of the 'give and take' kind. He will submit to the introduction of a trifling defect, in order that he may profit by a great excellence; and between excellencies, perhaps somewhat incompatible, he will decide on which is the greatest, and give it the preference.

"To a person commencing improvement, the best advice is to get as good a *bull* as he can; and if he be a good one of his kind, to use him indiscriminately with all his cows; and when by this proceeding, which ought to be persisted in, his stock has, with an occasional change of bull, become sufficiently stamped with desirable excellencies, his selection of males should then be made, to eradicate defects which he thinks it desirable to get rid of.

"He will not fail to keep in view the necessity of *good blood* in the bulls resorted to, for that will give the only *assurance* that they will transmit their own valuable properties to their offspring; but he must not depend on this alone, or he will soon run the risk of degeneracy.

"In animals evincing an extraordinary degree of perfection, and where the constitution is decidedly good, and there is no prominent defect, a little close breeding may be allowed—as the son with the mother, to whom he is only half-blood—or the brother with the sister. But this must not be injudiciously adopted or carried too far, for although it may increase and confirm valuable properties, it will also increase and confirm defects; and no breeder need be long in discovering that in an improved state animals have a greater tendency to defect than to perfection. Close breeding, from affinities, impairs the constitution, and affects the procreative powers, and therefore a strong cross is occasionally necessary."

† When heifers of this age will not stand their bulling, a couple of doses of physic, or the turning on shorter pasture until they next come into season, will set all right.

Mr. Parkinson's opinion, although somewhat different in one point from that we have stated, deserves consideration:—"I had three heifers, when I lived at Slane,

[*Editor's Note:* Material has been omitted at this point.]

took the bull at one year old, I believe, in consequence of their being reared in the open air at the haystacks, which caused them to be forwarder. I had not the least idea of this happening, or I should have prevented it, as I think it very injudicious. It is the opinion of some persons, that by suffering heifers to be three or four years old they make fine cattle, but I never found any material difference; while there is a loss of one year, besides the danger of not standing the bulling; and it adds very much to the profit of the heifer if she be given to the bull at two or two-and-a-half years old, for the time she is in calf, added to that of the calf sucking and the time she will be fattening, bring her to four or four years and a half when she is slaughtered. A heifer that has had a calf will fatten quicker and tallow better than one of the same age that has not, while a calf is gained, worth, if of a good breed, eight or ten pounds as a store beast."—*Treatise on Live Stock*, vol. i. p. 99.

12

How can Dr. Prichard say, there are no constitutional differences — smell of negro — like differences between [illegible] common Dog & the goats — Cashmir goat in Penny Magazine said not to smell

THE

ART

OF

IMPROVING THE BREEDS

OF

DOMESTIC ANIMALS.

IN A LETTER

ADDRESSED TO

THE

RIGHT HON. SIR JOSEPH BANKS, K. B.

BY

SIR JOHN SAUNDERS SEBRIGHT, BART. M.P.

LONDON:

PRINTED FOR JOHN HARDING,

36, ST. JAMES'S STREET.

[*Editor's Note:* The annotations are those of Charles Darwin. Some were made in pen, others in pencil; hence the variations in density.]

Reprinted from pages i and 3–31 of *The Art of Improving the Breeds of Domestic Animals in a Letter Addressed to the Right Hon. Sir Joseph Banks, K.B.*, J. Harding, London, 1809, 31pp.

A

LETTER,

&c. &c.

DEAR SIR,

I HAVE not the presumption to think, that I can throw any light upon the art of improving the breeds of domestic animals, which is now so well understood in this country: but in obedience to your commands, I print these observations, to which I am sensible you have attached more value than they deserve.

The attention which gentlemen of landed property have, of late years, paid to this subject, has been extremely beneficial to the country; not so much by the improvements which they themselves have made, as by the encouragement which the

A 2

professional breeders have received from their patronage and support, without which they could not have carried the breeds of cattle and sheep, to the perfection which many of them have now attained.

They have, likewise, been the means of making the best breeds known in every part of the kingdom, and of transporting them to districts, where it is not probable they would have been introduced, but through their agency.

The Duke of Bedford, Mr. Coke, and some few others, have not only been the liberal patrons of the professional breeders, but have themselves made great improvements in the breeds, to which their attention has been directed.

The same success has not, in general, attended gentlemen in this pursuit: the best breeds, after having been obtained by them at a great expense, too frequently

5

degenerate in their hands, from mismanagement. They conceive, that, if they have procured good males and good females, they have done all that is necessary to establish and to continue a good breed, but this is by no means the case.

Were I to define what is called the art of breeding, I should say, that it consisted in the selection of males and females, intended to breed together, in reference to each other's merits and defects.

It is not always by putting the best male to the best female, that the best produce will be obtained; for should they both have a tendency to the same defect, although in ever so slight a degree, it will in general preponderate so much in the produce, as to render it of little value.

A breed of animals may be said to be improved, when any desired quality has been increased by art, beyond what that quality was in the same breed, in a state of nature : the swiftness of the race-horse, the propensity to fatten in cattle, and the fine wool in sheep, are improvements which have been made in particular varieties of the species to which these animals belong. What has been produced by art, must be continued by the same means, for the most improved breeds will soon return to a state of nature, or perhaps defects will arise, which did not exist when the breed was in its natural state, unless the greatest attention is paid to the selection of the individuals who are to breed together.

We must observe the smallest tendency to imperfection in our stock, the moment it appears, so as to be able to counteract it, before it becomes a defect ; as a rope-dancer, to preserve his equilibrium, must correct the balance, before it is gone too far, and then not by such a motion, as will incline it too much to the opposite side.

The breeder's success will depend entirely upon the degree in which he may happen to possess this particular talent.

Regard should not only be paid to the qualities apparent in animals, selected for breeding, but to those which have prevailed in the race from which they are descended, as they will always shew themselves, sooner or later, in the progeny: it is for this reason that we should not breed from an animal, however excellent, unless we can ascertain it to be what is called *well bred;* that is, descended from a race of ancestors, who have, through several generations, possessed, in a high degree, the properties which it is our object to obtain.

The offspring of some animals is very unlike themselves; it is, therefore, a good precaution, to try the young males with a few females, the quality of whose produce has been already ascertained: by this means we shall know the sort of stock they get, and the description of females to which they are the best adapted.

8

If a breed cannot be improved, or even continued in the degree of perfection at which it has already arrived, but by breeding from individuals, so selected as to correct each other's defects, and by a judicious combination of their different properties, (a position, I believe, that will not be denied), it follows that animals must degenerate, by being long bred from the same family, without the intermixture of any other blood, or from being what is technically called, *bred in-and-in.*

Mr. Bakewell, who certainly threw more light upon the art of breeding than any of his predecessors, was the first, I believe, who asserted that a cross was unnecessary, and that animals would not degenerate, by being *bred in-and-in*, which was at that time the received opinion.

He said, you could but breed from the best. Of this there can be no doubt; but it is to be proved, how long the same family, *bred in-and-in*, will continue to be the best.

9

No one can deny the ability of Mr. Bakewell, in the art of which he may fairly be said to have been the inventor: but the mystery with which he is well known to have carried on every part of his business, and the various means which he employed to mislead the public, induce me not to give that weight to his assertions, which I should do to his real opinion, could it have been ascertained.

Mr. Meynel's fox-hounds are likewise quoted as an instance of the success of this practice; but, upon speaking to that gentleman upon the subject, I found that he did not attach the meaning that I do, to the term *in-and-in*. He said, that he frequently bred from the father and the daughter, and the mother and the son. This is not what I consider as breeding *in-and-in;* for the daughter is only half of the same blood as the father, and will probably partake, in a great degree, of the properties of the mother.

10

Mr. Meynel sometimes bred from brother and sister: this is certainly what may be called a *little close:* but should they both be very good, and, particularly, should the same defects not predominate in both, but the perfections of the one promise to correct in the produce the imperfections of the other, I do not think it objectionable: much further than this, the system of breeding from the same family cannot, in my opinion, be pursued with safety.

Mr. Bakewell had certainly the merit of destroying the absurd prejudice which formerly prevailed against breeding from animals, between whom there was any degree of relationship; had this opinion been universally acted upon, no one could have been said to be possessed of a particular breed, good or bad; for the produce of one year would have been dissimilar to that of another, and we should have availed ourselves but little of an animal

This shows the whole principle of making varieties

of superior merit, that we might have had the good fortune to possess.

The authorities of Mr. Bakewell, and of Mr. Meynel, being generally quoted, when this subject is discussed, I have stated, why I reject that of the former altogether, and that the latter, in point of fact, never fairly tried the experiment.

I do not find that any of the many advocates for breeding *in-and-in*, with whom I have conversed, have tried it to any extent; they say, that it is to perfect animals only that the practice applies, but the existence of a perfect animal is an hypothesis I cannot admit.

I do not believe, that there ever did exist an animal without some defect, in constitution, in form, or in some other essential quality; a tendency, at least, to the same imperfection, generally prevails in different degrees in the same family. By breeding *in-and-in*, this defect, how-

Shows the control [illegible] / Variation

ever small it may be at first, will increase in every succeeding generation; and will, at last, predominate to such a degree, as to render the breed of little value. Indeed, I have no doubt but that by this practice being continued, animals would, in course of time, degenerate to such a degree, as to become incapable of breeding at all.

The effect of *breeding in-and-in* may be accelerated, or retarded by selection, particularly in those animals who produce many young ones at a time. There may be families so nearly perfect, as to go through several generations, without sustaining much injury, from having been bred *in-and-in;* but a good judge would, upon examination, point out by what they must ultimately fail, as a mechanic would discover the weakest part of a machine, before it gave way.

Breeding *in-and-in*, will, of course, have the same effect in strengthening the good,

13

as the bad properties, and may be beneficial, if not carried too far, particularly in fixing any variety which may be thought valuable.

I have tried many experiments, by breeding *in-and-in* upon dogs, fowls, and pigeons: the dogs became, from strong spaniels, weak and diminutive lap-dogs, the fowls became long in the legs, small in the body, and bad breeders.

There are a great many sorts of fancy-pigeons; each variety has some particular property, which constitutes its supposed value, and which the amateurs increase as much as possible, both by breeding *in-and-in*, and by selection, until the particular property is made to predominate to such a degree, in some of the most refined sorts, that they cannot exist without the greatest care, and are incapable of rearing their young, without the assistance of other pigeons, kept for that purpose.

The Leicestershire breeders of sheep

have inherited the principles, as well as the stock, of their leader, Mr. Bakewell: he very properly considered a propensity to get fat, as the first quality in an animal, destined to be the food of man: his successors have carried this principle too far; their stock are become small in size, and tender, produce but little wool, and are bad breeders.

By selecting animals for one property only, the same effect will, in some degree, be produced, as by breeding *in-and-in*: we shall obtain animals, with the desired property in great perfection, but so deficient, in other respects, as to be upon the whole an unprofitable stock.

We should, therefore, endeavour to obtain all the properties that are essential to the animals we breed. The Leicestershire sheep prove that too much may be sacrificed, even to that most desirable quality in grazing stock—a disposition to get fat at an early age, and with a small quantity of food.

In plants man prevents mixture varieties cultivation & destroys, the unformed kinds — could he do the last *effectually* & keep on same land whether for *many* generations he would [illegible]

15

Many causes combine to prevent animals, in a state of nature, from degenerating; they are perpetually intermixing, and therefore do not feel the bad effects of breeding *in-and-in*: the perfections of some correct the imperfections of others, and they go on without any material alteration, except what arises from the effects of food and climate.

The greatest number of females will, of course, fall to the share of the most vigorous males; and the strongest individuals of both sexes, by driving away the weakest, will enjoy the best food, and the most favourable situations, for themselves and for their offspring.

A severe winter, or a scarcity of food, by destroying the weak and the unhealthy, has all the good effects of the most skilful selection. In cold and barren countries no animals can live to the age of maturity, but those who have strong constitutions; the weak and the unhealthy do

which would be infertile unless the selection —

16

not live to propagate their infirmities, as is too often the case with our domestic animals. To this I attribute the peculiar hardiness of the horses, cattle, and sheep, bred in mountainous countries, more than to their having been inured to the severity of the climate; for our domestic animals do not become more hardy by being exposed, when young, to cold and hunger: animals so treated will not, when arrived at the age of maturity, endure so much hardship as those who have been better kept in their infant state.

If one male, and one female only, of a valuable breed, could be obtained, the offspring should be separated, and placed in situations as dissimular as possible; for animals kept together are all subjected to the effects of the same climate, of the same food, and of the same mode of treatment, and consequently to the same diseases, particularly to such as are infectious, which must accelerate the bad effects of *breeding in-and-in.*

17

By establishing the breed in different places, and by selecting, with a view to obtain different properties in these several colonies, we may perhaps be enabled to continue the breed for some time, without the intermixture of other blood.

If the original male and female were of different families, by breeding from the mother and the son, and again from the male produce and the mother, and from the father and the daughter in the same way, two families sufficiently distinct might be obtained; for the son is only half of the father's blood, and the produce from the mother and the son will be six parts of the mother and two of the father.

Although I believe the occasional intermixture of different families to be necessary, I do not, by any means, approve of mixing two distinct breeds, with the view of uniting the valuable properties of both: this experiment has been frequently tried by others, as well as by myself, but has, I believe never succeeded. The first cross frequently produces a tolerable animal, but it is a breed that cannot be continued.

If it were possible, by a cross between the new Leicestershire and Merino breeds of sheep, to produce an animal uniting the excellencies of both, that is, the carcase of the one with the fleece of the other, even such an animal so produced would be of little value to the breeder; a race of the same description could not be perpetuated; and no dependance could be placed upon the produce of such animals; they would be mongrels, some like the new Leicester, some like the Merino, and most of them with the faults of both.

Merino rams are frequently put to South Down and Ryeland ewes, not with a view of obtaining the good properties of both kinds, but from the difficulty of procuring

Spanish ewes, and with the intention of obtaining the Merino blood in sufficient purity, for every practical purpose, by repeatedly crossing the female produce with Merino males.

I have no doubt but that better stock may be obtained, in a few years, in this manner, from a large flock of well-chosen ewes, than by breeding, at first, from a small number of the pure Merino blood, (and many of them cannot be obtained;) for the great advantage to be derived from the means of selection afforded by a more numerous flock, will more than compensate for the little stain of impure blood, which would be insensible in a flock, crossed in this manner, for four or five generations.

The introduction of Merino sheep to this country opens a fine field for improvement: it has been ascertained, that neither the sheep nor the wool sustain any injury from the change of climate or pas-

B 2

ture; and the absurd prejudice, that Merino wool could be grown only in Spain, is fortunately eradicated.

In comparing the Merino sheep with the South Downs, which are allowed to be the best of our short-woolled breeds, the former have very much the advantage, both as to quantity and quality of wool; but, I believe, the latter would produce by far the greatest quantity of meat, from a given quantity of food, which is the criterion by which we determine the relative value of all animals as grazier's stock.

Taking the gross produce, both of wool and of carcase, at the present prices, the Merino breed may perhaps be the most profitable: but should it be generally introduced, fine wool would become cheaper, and mutton dearer; it is therefore not easy to form a conclusive opinion upon this subject.

Great improvements may undoubtedly

be made in the Merino breed, as to their disposition to get fat. Their advocates say, with truth, that the South Down sheep were but a few years ago as imperfect in shape as the Merino now are; but they should recollect, that a disposition to fatten at an early age was always the characteristic of the South Down breed, even in its most unimproved state, and that it was from its possessing this very essential quality that so much attention has been paid to it.

It is well known that a particular formation generally indicates a disposition to get fat, in all sorts of animals; but this rule is not universal, for we sometimes see animals of the most approved forms, who are *slow feeders*, and whose flesh is of a bad quality, which the graziers easily ascertain by the *touch*. The disposition to get fat is more generally found in some breeds than in others. The Scotch Highland cattle are remarkable for being almost all *quick feeders*, although many of them are defective in shape. The Welsh cattle have but little disposition to get fat: not from being particularly ill-shaped, but because they are almost invariably what the graziers call *bad handlers*.

We must not therefore suppose, that the bad shape of the Merino sheep is the sole cause of its being so ill calculated for the purpose of the grazier.

An observation which Dr. Jenner made to me about ten years ago, (the truth of which has since been confirmed by my own experience—that no animal whose chest was narrow could easily be made fat,) applies particularly to the Merino sheep, who are in general contracted in that part, and is well worth the attention of those who wish to improve this breed.

Perhaps the great secretion of yolk, so essential to the production of fine wool, and which is excessive in the Merino sheep, may be incompatible with the fattening quality.

I have always found the fineness of the fleece in exact proportion to the quantity of yolk it contained. Those who are unaccustomed to examine wool, may consider this as a certain criterion of its quality: for although the hair of some dry fleeces may be fine, it will always want the elasticity which is so much valued by the manufacturer.

It is to be regretted, that so little attention has been paid to the improvement of British wool, and particularly to that of the short-woolled breeds: a fine fleece is not only more profitable to the owner, but from the closeness of its texture, and the quantity of yolk it always contains, is as much better protection to the sheep in bad weather, than the open and hairy covering, which too generally disgraces our flocks.

This extraordinary negligence in the sheep-breeders may, in some degree, be accounted for, by the manner in which the wool-trade is carried on. The *growers* are seldom well acquainted with the value of this article, or indeed with its quality, and the *buyers* find their account in fixing a general price every year, for the wool of each breed, without making any distinction between the very different quality of the pile of different flocks.

It is likewise the custom, in many parts of England, for the growers to deliver their wool to the buyers, upon their engaging to give them the highest price of the year; such bargains (and they are very general) are of course strong inducements to the purchasers not to give the full value for any wool that may be offered to them of a superior quality.

The fineness of the fleece, like every other property in animals of all kinds, may be improved by selection in breeding. The opinion, that good wool could only be produced in particular districts, is a prejudice which fortunately no longer exists.

Climate, food, and soil, have certainly some effect upon the quality of wool, but not so much as is generally supposed. The fleece is affected by the degree of nourishment which the animal receives, not by the quality of the pasture on which it is fed. If sheep are highly kept, their wool will become less fine, but in other respects its quality will not be deteriorated. The wool of a starved sheep may be apparently fine, but it will be brittle, and of little value to the manufacturer.

A regular supply of food to the sheep is essential to the growth of good wool; for that part of the hair which grows when the animal is in a high state of flesh, will be thick, and that which is grown when it is reduced by hunger, will be weak and thin; and consequently the thickness of hair will always be irregular, if the animal passes from one extreme to the other.

The alteration which may be made in any breed of animals by selection, can hardly be conceived by those who have not paid some attention to this subject; they attribute every improvement to a cross, when it is merely the effect of judicious selection.

I have often been told, that from the beautiful shape of Mr. Elman's South Down sheep, they must have been crossed with the New Leicester; and that from the fineness of their wool, they must have been crossed with the Merino breed; but I do not conceive, that even the skill of this very distinguished breeder could have retained the good shape of the former, without any appearance of the coarseness of its wool, or the fine fleece of the latter, without the deformity of its carcase, had he crossed his flock with either of these breeds.

It may as well be contended, that the white pheasant, which is now become very common, was produced from a cross with a Dorking fowl, whereas it was one of those accidental varieties which sometimes occur, and which has been perpetuated by selection. The same may be said of the endless variety in the colour, shape, and size, of rabbits, ducks, and pigeons, in a domesticated state; a variety produced by the art of man, and which did not exist in these creatures in their natural state.

A greater proof, I conceive, of what may be effected by selection and perseverance, cannot be adduced.

There is, perhaps, no means by which the breeds of animals can be so rapidly, and so effectually improved, as by its being the particular business of some breeders to provide male animals for the purpose of letting for hire. Our horses could never have arrived at the degree of perfection which they have now attained, but from the facility which has been afforded to every one, by the public stallions, of breeding from the best horses of every description, at a moderate expense.

The breeds of sheep to which this practice has been applied, have attained great perfection, while those which have never been attended to by persons in this particular business, shew no signs of improvement.

No trouble or expense will be spared by those who expect to derive profit, not from the quantity, but from the quality of the animals which they breed. The competition, which must always exist between breeders of this description, will be a never-failing stimulus to exertion.

29

The common farmer, who seldom sees any stock but his own and that of his neighbours, generally concludes, that his own have arrived at the summit of perfection: but the breeder who lets for hire, must frequently submit his male animals to the inspection of the public, and to the criticism of his rivals, who will certainly not encourage any prejudices he may entertain of their superiority.

In this trade, as in every other, there ought to be a regular gradation; those, for example, who hire a male for eighty guineas, will be amply repaid by letting seven, or eight, for twenty guineas each, as will those who hire for twenty guineas, by letting several for five or six.

Thus each, besides the improvement of his stock, will receive a fair remuneration, and every breeder have the means of selecting the male he thinks best calculated for the females he may happen to possess.

30

The same effect will not be produced by the sale of male animals; for we are induced to keep a male we have purchased at a high price, although we may not be entirely satisfied with his produce, but by hiring, we endeavour to select a male every year, with the properties in which our females are deficient, and whom we think calculated to correct the faults which arise from time to time in our stock.

These observations are the result of many years experience, in breeding animals of various descriptions: but the life of man is not long enough to form very decisive conclusions upon a subject which is so little understood, and which is darkened by innumerable prejudices. Many experiments must be tried, to establish a single fact; for Nature is sometimes so capricious in her productions, that the most accurate observer will be frequently deceived, if he draws any inference from a single experiment.

I have freely stated my opinions, without considering them as conclusive, and shall be much gratified if they induce others to direct their attention to a subject which appears to me of great importance to the agricultural interests of this country.

I have the honour to be,
DEAR SIR,
With great regard,
Your obedient humble Servant,

J. S. SEBRIGHT.

Beechwood, Aug. 1, 1809.

Howlett and Brimmer,
Printers, 10, Frith Street, Soho.

13

Reprinted from pages 2–5 of *Remarks on the Improvement of Cattle, &c. In a Letter to Sir John Saunders Sebright, Bart. M.D.*, 3rd ed., H. Barnett, Nottingham, 1820

REMARKS ON THE IMPROVEMENT OF CATTLE, &c. IN A LETTER TO SIR JOHN SAUNDERS SEBRIGHT, BART. M.D.

J. Wilkinson

[*Editor's Note:* In the original, material precedes this excerpt.]

It has always occurred to me, that in order to understand the true art of breeding, we must have recourse to first principles; and that it will be easier to shew how any improved breed may be continued, when we have previously shewn how it was first formed.

In the following remarks, I shall confine myself chiefly to Neat Cattle, as being that species of animal with which I am more particularly concerned; but it will be easy to perceive that the observations there advanced, will be applicable, in a greater or less degree, to every other kind which is destined to be the food of man.

Whether the different breeds with which we are now acquainted, descended originally from one common stock, the wild Bison, is a question, I think, hard to be determined. Of this, however, we may be assured, from the very nature of the case, that the distinct breeds at first, if more than one, could have been by no means numerous; so that the great variety which we behold at present, is owing to food, to climate, or to other collateral and accidental circumstances. And perhaps of all the causes contributing to this multiplicity, none would be more effectual, than the hidden springs of nature itself. For though we perceive that there is a *strong* tendency, for *like to produce like,* as it is usually termed; yet he that is at all conversant with nature, must perceive also, that there is a *certain* tendency to change. And this law of nature would soon be assisted by man, who is ever fond of novelty; and delights in diversity, even for its own sake.

Thus then, we have seen, that distinct breeds might readily be formed by the joint efforts of nature and of art; nor will it be more difficult to perceive how they might afterwards be improved. That all would be capable of improvement is too obvious to need discussion. For no one can behold any breed whatever in its more natural and less improved state, without perceiving a great variety in the shapes of individuals, their different degrees of tendency to feeding, or certain other remarkable properties, which might give to some a decided superiority over the rest. These, therefore, must be selected from the whole herd; and as you yourself, Sir, have remarked, the male and female be properly matched. When we come to their progeny, some will probably be worse, some equal to, and some even better, than the parents themselves. The worst must unquestionably be rejected, while the rest, and especially the best of these, are carefully to be preserved for future stock. And thus by a judicious selection of male and female, and discarding every thing that is refuse, we must continue to proceed. And by such procedure, animals have at length been produced, so different from the generality of the stock from whence they were originally taken, that none but such as are well acquainted with these matters, could have any idea, that there existed between them the least affinity The distinction indeed between some, and their own particular variety, has scarcely been less, than the distinction between that variety and the whole species. The longer also these perfections have been continued, the more stability will they have acquired, and the more will they partake of nature itself. As to the leading properties which may constitute the excellence of any breed, or of any particular family belonging to that breed, I shall next inquire.

And first with respect to form; in which case, I shall give, what I conceive to be the most important points for the true symmetry of Neat Cattle in general. These are as follows.

The head ought to be rather long, and muzzle fine; the countenance calm and placid, which indicates a disposition to get fat; the horns fine; the neck light, particularly where it joins the head; the breast wide and projecting well before the legs; the shoulders moderately broad at the top and the points well in, and when the animal is in good condition, the chine so full

[*Editor's Note:* Material has been omitted at this point.]

Part III

ARTIFICIAL SELECTION AND DARWIN'S ADVOCACY OF THE THEORY OF ADAPTIVE EVOLUTION BY SELECTION

Editor's Comments on Papers 14 and 15

14 DARWIN
Excerpt from *Variation under Domestication*

15 DARWIN
Excerpts from *Selection by Man*

Charles Darwin, in all six editions of his *On the Origin of Species*, contended that three types of selection were involved in bringing about "descent with modification" (evolution). Darwin divided the ecological interactions an organism has that cause evolution into the following three categories: "man's power of selection" (artificial selection)*, natural selection, and sexual selection (Figure 1). By dividing selection into artificial, natural, and sexual selection, Darwin was able, first, to argue by analogy from artificial selection, a process already widely known to bring about hereditary changes in domesticated species, to both natural and sexual selection, which were operating in the rest of nature (Figure 2 and 3); and second, to partition those intraspecific ecological interactions that selected for adaptations that were advantageous to the individuals possessing them but harmful to certain other members of the species into a separate selective system—sexual selection.

Charles Darwin divided artificial selection into unconscious selection and methodical selection in an attempt to bring about a better understanding of the power of man's selection in both *On the Origin of Species* (Darwin, 1859:34; see also Paper 14 and *Variation of Animals and Plants under Domestication* (Darwin, 1868, I:14, 234–235; see also Paper 15). He contrasted these two

*"Artificial selection" is a phrase that Darwin rarely used in his publications. He preferred to write about "man's power of selection." Darwin used the phrase "artificial selection" for the first time in a June 25, 1858, letter to Charles Lyell (Darwin, 1899, I: 474).

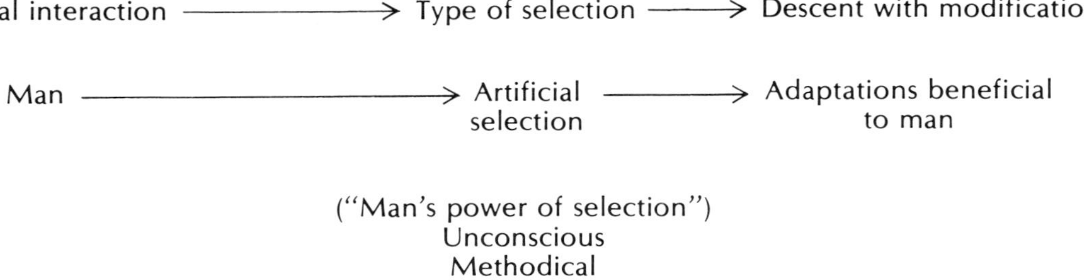
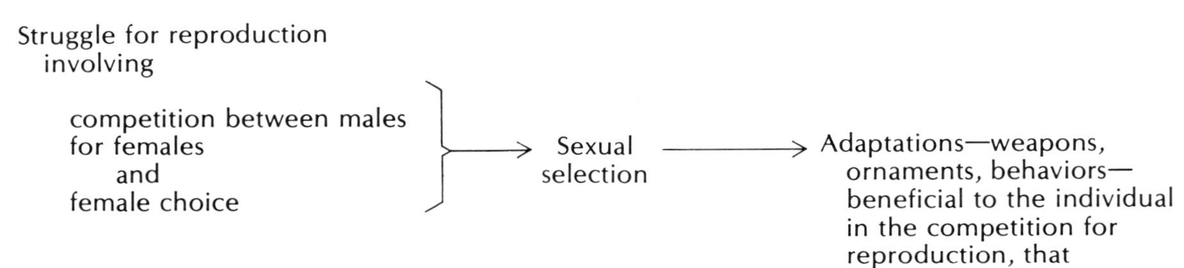

Figure 1. Charles Darwin's 1859 system for classifying selection. (After G. Hardin and C. Bajema, *Biology: Its Principles and Implications,* 3rd ed. San Francisco: W. H. Freeman, 1978.)

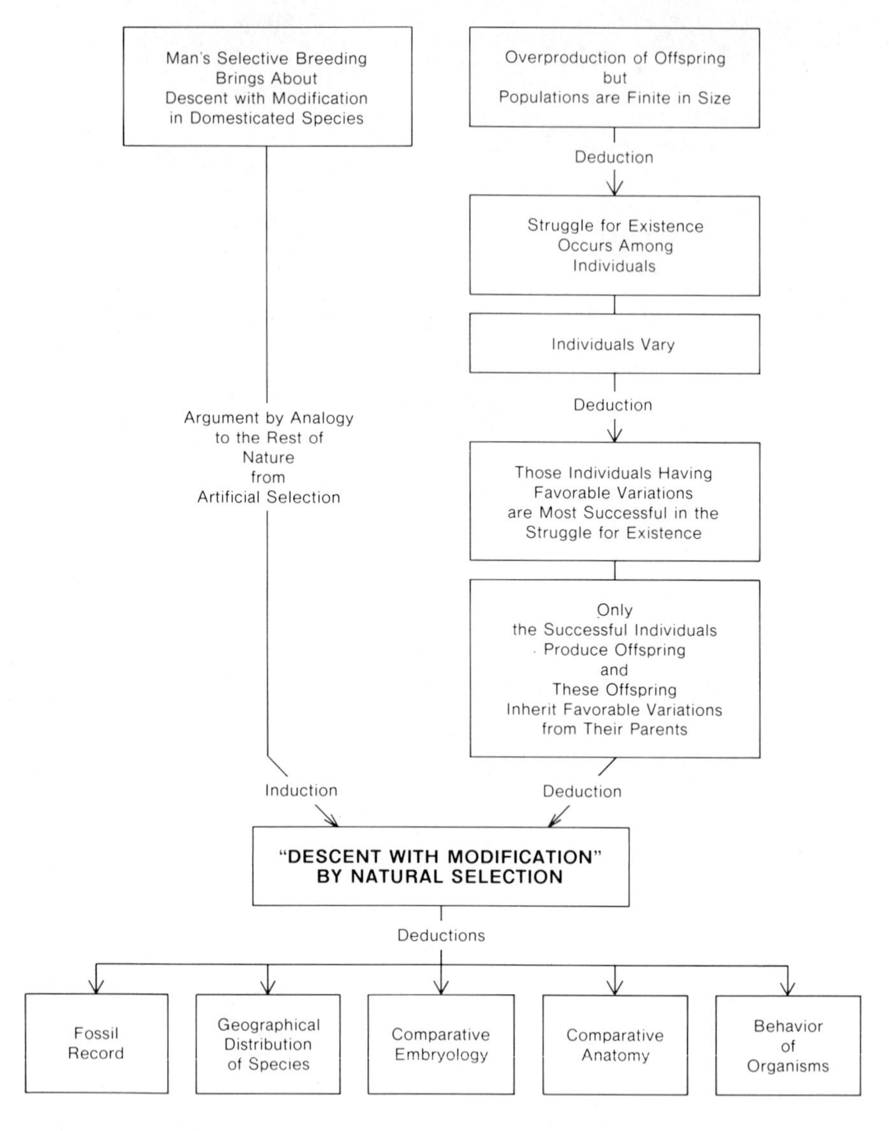

Figure 2. Darwin used three lines of reasoning in developing and advocating his theory of natural selection: the inductive vera causa ("true cause") argument from artificial selection; the deduction of the set of causes that produce natural selection; and the rationalist vera causa used to deduce the kinds of observations that would be made in nature if natural selection operated the way that Darwin predicted. (After M. Ruse, *The Darwinian Revolution: Science Red Tooth and Claw,* Chicago: University of Chicago Press, 1979, p. 198.)

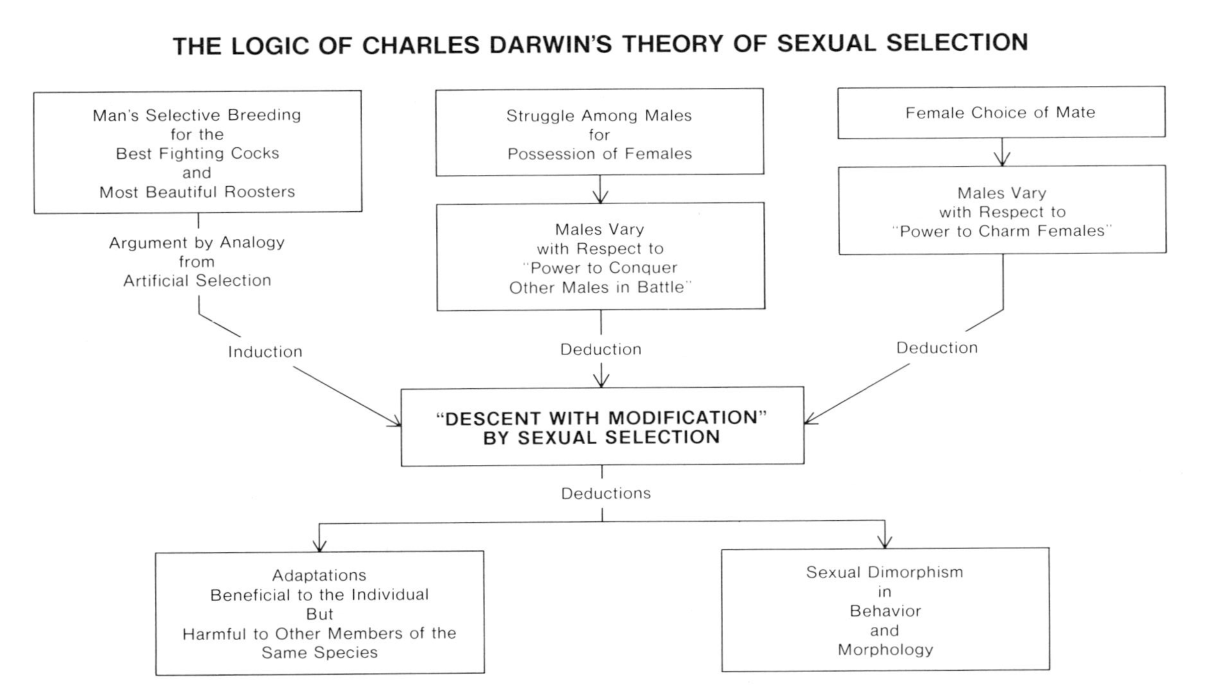

Figure 3. Darwin also used three lines of reasoning in developing and advocating his theory of sexual selection: the inductive vera causa ("true cause") argument from artificial selection; the deduction of the set of natural causes that produce sexual selection; and the rationalist vera causa used to deduce the kinds of observations that would be made in nature if sexual selection operated the way that Darwin predicted. (After M. Ruse, *The Darwinian Revolution: Science Red Tooth and Claw*, Chicago: University of Chicago Press, 1979, p. 198.)

ways by which humans cause domesticated plant and animal populations to undergo "descent with modification" as follows:

> Selection may be followed either methodically and intentionally, or unconsciously and unintentionally. Man may select and preserve each successive variation, with the distinct intention of improving and altering a breed, in accordance with a preconceived idea; and by thus adding up variations, often so slight as to be imperceptible by the uneducated eye, he has effected wonderful changes and improvements. It can, also, be clearly shown that man, without any intention or thought of improving the breed, by preserving in each successive generation the individuals he prizes most, and by destroying the worthless individuals, slowly, though surely, induces great changes. As the will of man thus comes into play, we can understand how it is that domesticated breeds show adaptation to his wants and pleasures. We can further understand how it is that domesticated races of animals and cultivated races of plants often exhibit an abnormal character, as compared with natural species; for they have been modified not for their own benefit, but for that of man. (Darwin, 1868, I:3–4).

Darwin contended that unconscious selection played a crucial role in the evolution of domesticated plants and animals in both the *Origin* and *Variation* (see Papers 14 and 15). Darwin also pointed out that natural selection operates on domesticated plants and animals often in direct opposition to the artificial selection being generated by man.

Darwin contended that natural selection operated almost exclusively on slight individual differences in the production of new species:

> The power of Selection, whether exercised by man, or brought into play under nature through the struggle for existence and the consequent survival of the fittest, absolutely depends on the variability of organic beings. Without variability nothing can be effected: slight individual differences, however suffice for the work, and are probably the sole differences which are effective in the production of new species. (Darwin, 1868, II:192)

Darwin's views on continuous versus discontinuous variation and on which type of variation was most important in evolution by selection have been analyzed by Osborn (1912), Ghiselin, (1969), Vorzimmer (1970), and Bowler (1974). Darwin's contention that sports (large discontinuous variations) played a minor role in evolution by selection became the center of a bitter controversy over the limits to selection during the first two decades of the twentieth century (see Part IV of this volume).

The argument that selection operating in nature was analogous to "man's power of selection" was considered by Charles Darwin to be one of the strongest arguments in favor of his theory of evolution by selection (Darwin, 1859, 1868, and in F. Darwin, 1899, II:209; Ruse, 1973). Darwin devoted the first chapter of *On the Origin of Species* to a discussion of variation under domestication and artificial selection (excerpted as Paper 14).* The following three passages taken from *Origin* demonstrate how Charles Darwin argued by analogy to natural selection from artificial selection:

> I have called this principle, by which each slight variation, if useful, is preserved, by the term of Natural Selection, in order to mark its relation to man's power of selection. We have seen that man by selection can certainly produce great results, and can adapt organic beings to his own uses, through the accumulation of slight but useful variations, given to him by the hand of Nature. But Natural Selection, as we shall hereafter see, is a power incessantly ready for action, and is as immeasurably superior to man's feeble efforts, as the works of Nature are to those of Art. (Darwin, 1859:61)

> As man can produce and certainly has produced a great result by his methodical and unconscious means of selection, what may not nature effect? Man can act only on external and visible characters: nature cares nothing for appearances, except in so far as they may be useful to any being. She can act on every internal organ, on every shade of constitutional difference, on the whole machinery of life. Man selects only for his own good; Nature only for that of the being which she tends. Every selected character is fully exercised by her; and the being is placed under well-suited conditions of life. Man keeps the natives of many climates in the same country; he seldom exercises each selected character in some peculiar and fitting manner; he feeds a long and a short beaked pigeon on the same food; he does not exercise a long-backed or long-legged quadruped in any peculiar manner; he exposes sheep with long and short wool to the same climate. He does not allow the most vigorous males to struggle for the females. He does not rigidly destroy all inferior animals, but protects during each varying season, as far as lies in his power, all his productions. He often begins his selection by some half-monstrous form; or at least by some modification prominent enough to catch his eye, or to be plainly useful to him. Under nature, the slightest difference of structure or constitution may well turn the nicely-balanced scale in the struggle for life, and so be preserved. How fleeting are the wishes and efforts of man! how short his time! and consequently how poor will his products

*All the changes in text that Charles Darwin made in the second and later editions are catalogued by Morse Peckham, editor, in *The Origin of Species by Charles Darwin, A Variorum Text* (Philadelphia: University of Pennsylvania Press, 1959), 816pp.

> be, compared with those accumulated by nature during whole geological periods. Can we wonder, then, that nature's productions should be far "truer" in character than man's productions; that they should be infinitely better adapted to the most complex conditions of life, and should plainly bear the stamp of far higher workmanship? (Darwin, 1859:83–84)
>
> Why, if man can by patience select variations most useful to himself, should nature fail in selecting variations useful, under changing conditions of life, to her living products? What limit can be put to this power, acting during long ages and rigidly scrutinising the whole constitution, structure, and habits of each creature—favouring the good and rejecting the bad? I can see no limit to this power, in slowly and beautifully adapting each form to the most complex relations of life. The theory of natural selection, even if we looked no further than this, seems to me to be in itself probable. (Darwin, 1859:469)

Darwin contributed to the development of the erroneous belief that "Natural Selection requires the constant watching of an intelligent 'chooser'" by choosing the term natural selection, by comparing its effects with man's [artificial] selection and by "so frequently personifying nature as 'selecting,' as 'preferring,' 'seeking only the good of the species." (Wallace, 1866). Wallace, in his July 2, 1866, letter to Charles Darwin, urged Darwin to adopt the metaphor "survival of the fittest" (which the philosopher Herbert Spencer had coined in 1864) to describe what Darwin had been calling "natural selection." Darwin first used "survival of the fittest" to describe natural selection in *Variation of Animals and Plants under Domestication* (Darwin, 1868:17). Spencer's metaphor was also used by Darwin to describe natural selection in the fifth (1869) and sixth (1872) editions of *On the Origin of Species*. The history of the confusion brought about by the use of Spencer's "survival of the fittest" metaphor to describe natural selection will be reviewed by Bajema in a forthcoming Benchmark volume on natural selection.

Darwin's strategy of using the analogy between artificial and natural selection to help others gain a better understanding of natural selection was repeatedly employed by evolutionary biologists after 1859. Thomas Huxley (1860:308) wrote:

> The Darwinian hypothesis has the merit of being eminently simple and comprehensible in principle, and its essential positions may be stated in very few words: all species have been produced by the development of varieties from common stocks; by the conversion of these, first into permanent races and then into new species by the process of *natural selection*, which process is essentially identical with that artificial selection by which man has originated the races of domestic

> animals—the *struggle for existence* taking the place of man, and exerting, in the case of natural selection, that selective action which he performs in artificial selection.

Huxley (1860, 1863) contended that the argument by analogy from artificial selection to natural selection was incomplete when the argument was used to explain the origin of species. While man has produced many "morphological species" (what we now call varieties or races) by artificial selection, Huxley pointed out that man had not yet produced any "physiological species," i.e., populations that are reproductively isolated from each other. Huxley (1860, 1863) urged:

> Mr. Darwin . . . to place his views beyond the reach of all possible assault . . . [by demonstrating] the possibility of developing from a particular stock by selective breeding, two forms, which should either be unable to cross one with another, or whose cross-bred offspring should be infertile with another . . . at present, so far as experiments have gone, it has not been found possible to produce this complete physiological divergence by selective breeding . . . if it could be proved, not only that this *has* not been done, but that it *cannot* be done; if it could be demonstrated that it is impossible to breed selectively, from any stock, a form which shall not breed with another produced from the same stock; and if we were shown that this must be the necessary and inevitable results of all experiments, I hold that Mr. Darwin's hypothesis would be utterly shattered. (Huxley, 1896:463–464)

The Benchmark volumes entitled the *Genetics of Speciation* (Jameson 1977*a*) and *Hybridization: An Evolutionary Perspective* (Levin 1979) both include reprinted papers that demonstrate the extent to which reproductive isolation has been produced in artificial and natural selection experiments.

Alfred Russel Wallace, in his preface to *Darwinism*, contended, "It has always been considered a weakness in Darwin's work that he based his theory, primarily, on the evidence of variation in domesticated animals and cultivated plants" rather than on the evidence for variation of organisms in a state of nature (Wallace, 1889:vi). Wallace nonetheless used the argument by analogy from artificial selection when he introduced natural selection to the readers of *Darwinism* (Wallace, 1889:102–103). The following succinct summary of the analogy between artificial and natural selection was presented by the German biologist August Weismann at the fiftieth anniversary celebration of the publication of *On the Origin of Species:* "Natural selection depends on the same three factors as artificial selection: on variability, inheritance, and selection for breeding, but this last is here carried out not by a breeder but by what Darwin called the 'struggle for existence'" (Weismann, 1909:19).

Charles Darwin also argued by analogy from artificial selection to sexual selection in *On the Origin of Species:*

> Sexual selection by always allowing the victor to breed might surely give indomitable courage, length to the spur, and strength to the wing to strike in the spurred leg, as well as the brutal cockfighter, who knows well that he can improve his breed by careful selection of the best cocks. . . .
> If man can in a short time give elegant carriage and beauty to his bantams, according to his standard of beauty, I can see no good reason to doubt that female birds, by selecting, during thousands of generations, the most melodious or beautiful males, according to their standard of beauty, might produce a marked effect. (Darwin, 1859:88–89)

Darwin's summary of his theory of sexual selection in the concluding chapter of *The Descent of Man and Selection in Relation to Sex* includes his argument by analogy:

> Sexual selection depends on the success of certain individuals over others of the same sex in relation to the propagation of the species; whilst natural selection depends on the success of both sexes, at all ages, in relation to the general conditions of life. The sexual struggle is of two kinds; in the one it is between the individuals of the same sex, generally the male sex, in order to drive away or kill their rivals, the females remaining passive; whilst in the other, the struggle is likewise between the individuals of the same sex, in order to excite or charm those of the opposite sex, generally the females, which no longer remain passive, but select the more agreeable partners. This latter kind of selection is closely analogous to that which man unintentionally, yet effectually, brings to bear on his domesticated productions, when he continues for a long time choosing the most pleasing or useful individuals, without any wish to modify the breed. (Darwin, 1871, II:398)

The role that the theory of sexual selection played in the historical development of Charles Darwin's theory of adaptive evolution by selection will be reviewed by Bajema in a forthcoming Benchmark volume on sexual selection.

Charles Darwin perceived that artificial selection of plants and animals under domestication plays the role of experiment in the scientific analysis of evolution. (Schweber, 1977). In his introduction to *The Variation of Animals and Plants under Domestication*, Darwin (1868, I:3) wrote:

> But the initial variation on which man works, and without which he can do nothing, is caused by slight changes in the condi-

> tions of life, which must have often occurred under nature. Man, therefore, may be said to have been trying an experiment on a gigantic scale; and it is an experiment which nature during the long lapse of time has incessantly tried. Hence it follows that the principles of domestication are important for us. The main result is that organic beings thus treated have varied largely, and the variations have been inherited.

Both of the times Charles Darwin contended that traits could evolve by what we now call kin selection (Hamilton, 1964, Maynard Smith, 1964) Darwin employed the argument by analogy from artificial selection (Darwin, 1859:237–238, 1871:161). Just how could traits that lowered the fertility of individuals expressing them be favored by selection? Darwin argued in *Origin* that:

> selection may be applied to the family, as well as to the individual, and may thus gain the desired end. Thus, a well-flavoured vegetable is cooked, and the individual is destroyed: but the horticulturalist sows seeds of the same stock, and confidently expects to get nearly the same variety: breeders of cattle wish the flesh and fat to be well marbled together: the animal has been slaughtered, but the breeder goes with confidence to the same family. I have such faith in the powers of selection, that I do not doubt that a breed of cattle, always yielding oxen with extraordinarily long horns, could be slowly formed by carefully watching which individual bulls and cows, when matched, produced oxen with the longest horns: and yet no one ox could ever have propagated its kind. Thus I believe it has been with social insects: a slight modification of structure, or instinct, correlated with the sterile condition of certain members of the community, has been advantageous to the community: consequently the fertile males and females of the same community flourished, and transmitted to their fertile offspring a tendency to produce sterile members having the same modification. And I believe that this process has been repeated, until that prodigious amount of difference between the fertile and sterile females of the same species has been produced, which we see in many social insects. (Darwin, 1859:237–238)

Scientific experiments employing artificial selection to study evolution were rare until the beginning of the twentieth century when scientists began to propose more explicit theories concerning evolution and to collect and analyze their data in a more objective quantitative way. This occurred as the result of the development of statistical techniques to study heredity and evolution by Francis Galton, Karl Pearson, and other biometricians, and the rediscovery of Mendelian patterns of inheritance in 1900. Many of the successes and failures of practical breeders who employed artificial selection to bring about desired changes in the hereditary composition of domesticated species between 1859 and 1900 are chronicled in Verlot (1865), Darwin (1868), Hoffman (1869), Shirreff (1873), Rimpau (1877, 1891), Vilmorin (1886), Rümker (1889), Bailey (1906), de Vries (1901–1903, 1904, 1907), Newman (1912), Roberts (1929), and Stubbe (1972).

14

Reprinted from pages 29–43 of *On the Origin of Species by Means of Natural Selection, or the Preservation of Favoured Races in the Struggle for Life,* John Murray, London, 1859

VARIATION UNDER DOMESTICATION

C. R. Darwin

[*Editor's Note:* in the original, material precedes and follows this excerpt.]

Ask, as I have asked, a celebrated raiser of Hereford cattle, whether his cattle might not have descended from long-horns, and he will laugh you to scorn. I have never met a pigeon, or poultry, or duck, or rabbit fancier, who was not fully convinced that each main breed was descended from a distinct species. Van Mons, in his treatise on pears and apples, shows how utterly he disbelieves that the several sorts, for instance a Ribston-pippin or Codlin-apple, could ever have proceeded from the seeds of the same tree. Innumerable other examples could be given. The explanation, I think, is simple: from long-continued study they are strongly impressed with the differences between the several races; and though they well know that each race varies slightly, for they win their prizes by selecting such slight differences, yet they ignore all general arguments, and refuse to sum up in their minds slight differences accumulated during many successive generations. May not those naturalists who, knowing far less of the laws of inheritance than does the breeder, and knowing no more than he does of the intermediate links in the long lines of descent, yet admit that many of our domestic races have descended from the same parents—may they not learn a lesson of caution, when they deride the idea of species in a state of nature being lineal descendants of other species?

Selection.—Let us now briefly consider the steps by which domestic races have been produced, either from one or from several allied species. Some little effect may, perhaps, be attributed to the direct action of the external conditions of life, and some little to habit; but he would be a bold man who would account by such agencies for the differences of a dray and race horse, a greyhound and bloodhound, a carrier and tumbler pigeon. One of the most remarkable features in our domesticated races is that we see in them adaptation, not indeed to the animal's or plant's own good, but to man's use or fancy. Some variations useful to him have probably arisen suddenly, or by one step; many botanists, for instance, believe that the fuller's teazle, with its hooks, which cannot be rivalled by any mechanical contrivance, is only a variety of the wild Dipsacus; and this amount of change may have suddenly arisen in a seedling. So it has probably been with the turnspit dog; and this is known to have been the case with the ancon sheep. But when we compare the dray-horse and race-horse, the dromedary and camel, the various breeds of sheep fitted either for cultivated land or mountain pasture, with the wool of one breed good for one purpose, and that of another breed for another purpose; when we compare the many breeds of dogs, each good for man in very different ways; when we compare the game-cock, so pertinacious in battle, with other breeds so little quarrelsome, with "everlasting layers" which never desire to sit, and with the bantam so small and elegant; when we compare the host of agricultural, culinary, orchard, and flower-garden races of plants, most useful to man at different seasons and for different purposes, or so beautiful in his eyes, we must, I think, look further than to mere variability. We cannot suppose that all the breeds were suddenly produced as perfect and as useful as we now see them; indeed, in several cases, we know that this has not been their history. The key is man's power of accumulative selection: nature gives successive variations; man adds them up in certain directions useful to him. In this sense he may be said to make for himself useful breeds.

The great power of this principle of selection is not hypothetical. It is certain that several of our eminent breeders have, even within a single lifetime, modified to a large extent some breeds of cattle and sheep. In order fully to realise what they have done, it is almost necessary to read several of the many treatises devoted to this subject, and to inspect the animals. Breeders habitually speak of an animal's organisation as something quite plastic, which they can model almost as they please. If I had space I could quote numerous passages to this effect from highly competent authorities. Youatt, who was probably better acquainted with the works of agriculturists than almost any other individual, and who was himself a very good judge of an animal, speaks of the principle of selection as "that which enables the agriculturist, not only to modify the character of his flock, but to change it altogether. It is the magician's wand, by means of which he may summon into life whatever form and mould he pleases." Lord Somerville, speaking of what breeders have done for sheep, says:—"It would seem as if they had chalked out upon a wall a form perfect in itself, and then had given it existence." That most skilful breeder, Sir John Sebright, used to say, with respect to pigeons, that "he would produce any given feather in three years, but it would take him six years to obtain head and beak." In Saxony the importance of the principle of selection in regard to merino sheep is so fully recognised, that men follow it as a trade: the sheep are placed on a table and are studied, like a picture by a connoisseur; this is done three times at intervals of months, and the sheep are each time marked and classed, so that the very best may ultimately be selected for breeding.

What English breeders have actually effected is proved by the enormous prices given for animals with a good pedigree; and these have now been exported to almost every quarter of the world. The improvement is by no means generally due to crossing different breeds;

all the best breeders are strongly opposed to this practice, except sometimes amongst closely allied sub-breeds. And when a cross has been made, the closest selection is far more indispensable even than in ordinary cases. If selection consisted merely in separating some very distinct variety, and breeding from it, the principle would be so obvious as hardly to be worth notice; but its importance consists in the great effect produced by the accumulation in one direction, during successive generations, of differences absolutely inappreciable by an uneducated eye—differences which I for one have vainly attempted to appreciate. Not one man in a thousand has accuracy of eye and judgment sufficient to become an eminent breeder. If gifted with these qualities, and he studies his subject for years, and devotes his lifetime to it with indomitable perseverance, he will succeed, and may make great improvements; if he wants any of these qualities, he will assuredly fail. Few would readily believe in the natural capacity and years of practice requisite to become even a skilful pigeon-fancier.

The same principles are followed by horticulturists; but the variations are here often more abrupt. No one supposes that our choicest productions have been produced by a single variation from the aboriginal stock. We have proofs that this is not so in some cases, in which exact records have been kept; thus, to give a very trifling instance, the steadily-increasing size of the common gooseberry may be quoted. We see an astonishing improvement in many florists' flowers, when the flowers of the present day are compared with drawings made only twenty or thirty years ago. When a race of plants is once pretty well established, the seed-raisers do not pick out the best plants, but merely go over their seed-beds, and pull up the "rogues," as they call the plants that deviate from the proper standard. With animals this kind of selection is, in fact, also followed; for hardly any one is so careless as to allow his worst animals to breed.

In regard to plants, there is another means of observing the accumulated effects of selection—namely, by comparing the diversity of flowers in the different varieties of the same species in the flower-garden; the diversity of leaves, pods, or tubers, or whatever part is valued, in the kitchen-garden, in comparison with the flowers of the same varieties; and the diversity of fruit of the same species in the orchard, in comparison with the leaves and flowers of the same set of varieties. See how different the leaves of the cabbage are, and how extremely alike the flowers; how unlike the flowers of the heartsease are, and how alike the leaves; how much the fruit of the different kinds of gooseberries differ in size, colour, shape, and hairiness, and yet the flowers present very slight differences. It is not that the varieties which differ largely in some one point do not differ at all in other points; this is hardly ever, perhaps never, the case. The laws of correlation of growth, the importance of which should never be overlooked, will ensure some differences; but, as a general rule, I cannot doubt that the continued selection of slight variations, either in the leaves, the flowers, or the fruit, will produce races differing from each other chiefly in these characters.

It may be objected that the principle of selection has been reduced to methodical practice for scarcely more than three-quarters of a century; it has certainly been more attended to of late years, and many treatises have been published on the subject; and the result, I may add, has been, in a corresponding degree, rapid and important. But it is very far from true that the principle is a modern discovery. I could give several references to the full acknowledgment of the importance of the principle in works of high antiquity. In rude and barbarous periods of English history choice animals were often imported, and laws were passed to prevent their exportation: the destruction of horses under a certain size was ordered, and this may be compared to the "roguing" of plants by nurserymen. The principle of selection I find distinctly given in an ancient Chinese encyclopædia. Explicit rules are laid down by some of the Roman classical writers. From passages in Genesis, it is clear that the colour of domestic animals was at that early period attended to. Savages now sometimes cross their dogs with wild canine animals, to improve the breed, and they formerly did so, as is attested by passages in Pliny. The savages in South Africa match their draught cattle by colour, as do some of the Esquimaux their teams of dogs. Livingstone shows how much good domestic breeds are valued by the negroes of the interior of Africa who have not associated with Europeans. Some of these facts do not show actual selection, but they show that the breeding of domestic animals was carefully attended to in ancient times, and is now attended to by the lowest savages. It would, indeed, have been a strange fact, had attention not been paid to breeding, for the inheritance of good and bad qualities is so obvious.

At the present time, eminent breeders try by methodical selection, with a distinct object in view, to make a new strain or sub-breed, superior to anything existing in the country. But, for our purpose, a kind of Selection, which may be called Unconscious, and which results from every one trying to possess and breed from the best individual animals, is more important. Thus, a man who intends keeping pointers naturally tries to get as good dogs as he can, and afterwards breeds from his own best dogs, but he has no wish or expectation of permanently altering the breed. Nevertheless I cannot doubt that this process, continued during centuries, would improve and modify any breed, in the same way as Bakewell, Collins, &c., by this very same process, only carried on more methodically, did greatly modify, even during their own lifetimes, the forms and qualities of their cattle. Slow and insensible changes of this kind could never be recognised unless actual measurements or careful drawings of the breeds in question had been made long ago, which might serve for comparison. In some cases, however, unchanged or but little changed individuals of the same breed may be found in less civilised districts, where the breed has been less improved. There is reason to believe that King Charles's spaniel has been unconsciously modified to a large extent since the time of that monarch. Some highly competent authorities are convinced that the setter is directly derived from the spaniel, and has probably been slowly altered from it. It is known that the English pointer has been greatly changed within the last century, and in this case the change has, it is believed, been chiefly effected by crosses with the fox-hound; but what concerns us is, that the change has been effected unconsciously and gradually, and yet so effectually, that, though the old Spanish pointer certainly came from Spain, Mr. Borrow has not seen, as I am informed by him, any native dog in Spain like our pointer.

By a similar process of selection, and by careful training, the whole body of English racehorses have come to surpass in fleetness and size the parent Arab stock, so that the latter, by the regulations for the Goodwood Races, are favoured in the weights they carry. Lord Spencer and others have shown how the cattle of England have increased in weight and in early maturity, compared with the stock formerly kept in this country. By comparing the accounts given in old pigeon treatises of carriers and tumblers with these breeds as now existing in Britain, India, and Persia, we can, I think, clearly trace the stages through which they have insensibly passed, and come to differ so greatly from the rock-pigeon.

Youatt gives an excellent illustration of the effects of a course of selection, which may be considered as unconsciously followed, in so far that the breeders could never have expected or even have wished to have produced the result which ensued—namely, the production of two distinct strains. The two flocks of Leicester sheep kept by Mr. Buckley and Mr. Burgess, as Mr. Youatt remarks, "have been purely bred from the original stock of Mr. Bakewell for upwards of fifty years. There is not a suspicion existing in the mind of any one at all acquainted with the subject that the owner of either of them has deviated in any one instance from the pure blood of Mr. Bakewell's flock, and yet the difference between the sheep possessed by these two gentlemen is so great that they have the appearance of being quite different varieties."

If there exist savages so barbarous as never to think of the inherited character of the offspring of their domestic animals, yet any one animal particularly useful to them, for any special purpose, would be carefully preserved during famines and other accidents, to which savages are so liable, and such choice animals would thus generally leave more offspring than the inferior ones; so that in this case there would be a kind of unconscious selection going on. We see the value set on animals even by the barbarians of Tierra del Fuego, by their killing and devouring their old women, in times of dearth, as of less value than their dogs.

In plants the same gradual process of improvement, through the occasional preservation of the best individuals, whether or not sufficiently distinct to be ranked at their first appearance as distinct varieties, and whether or not two or more species or races have become blended together by crossing, may plainly be recognised in the increased size and beauty which we now see in the varieties of the heartsease, rose, pelargonium, dahlia, and other plants, when compared with the older varieties or with their parent-stocks. No one would ever expect to get a first-rate heartsease or dahlia from the seed of a wild plant. No one would expect to raise a first-rate melting pear from the seed of the wild pear, though he might succeed from a poor seedling growing wild, if it had come from a garden-stock. The pear, though cultivated in classical times, appears, from Pliny's description, to have been a fruit of very inferior quality. I have seen great surprise expressed in horticultural works at the wonderful skill of gardeners, in having produced such splendid results from such poor materials; but the art, I cannot doubt, has been simple, and, as far as the final result is concerned, has been followed almost unconsciously. It has consisted in always cultivating the best known variety, sowing its seeds, and, when a slightly better variety has chanced to appear, selecting it, and so onwards. But the gardeners of the classical period, who cultivated the best pear they could procure, never thought what splendid fruit we should eat; though we owe our excellent fruit, in some small degree, to their having naturally chosen and preserved the best varieties they could anywhere find.

A large amount of change in our cultivated plants, thus slowly and unconsciously accumulated, explains, as I believe, the well-known fact, that in a vast number of cases we cannot recognise, and therefore do not know, the wild parent-stocks of the plants which have been longest cultivated in our flower and kitchen gardens. If it has taken centuries or thousands of years to improve or modify most of our plants up to their present standard of usefulness to man, we can understand how it is that neither Australia, the Cape of Good Hope, nor any other region inhabited by quite uncivilised man, has afforded us a single plant worth culture. It is not that these countries, so rich in species, do not by a strange chance possess the aboriginal stocks of any useful plants, but that the native plants have not been improved by continued selection up to a standard of perfection comparable with that given to the plants in countries anciently civilised.

In regard to the domestic animals kept by uncivilised man, it should not be overlooked that they almost always have to struggle for their own food, at least during certain seasons. And in two countries very differently circumstanced, individuals of the same species, having slightly different constitutions or structure, would often succeed better in the one country than in the other, and thus by a process of "natural selection," as will hereafter be more fully explained, two sub-breeds might be formed. This, perhaps, partly explains what has been remarked by some authors, namely, that the varieties kept by savages have more of the character of species than the varieties kept in civilised countries.

On the view here given of the all-important part which selection by man has played, it becomes at once obvious, how it is that our domestic races show adaptation in their structure or in their habits to man's wants or fancies. We can, I think, further understand the frequently abnormal character of our domestic races, and likewise their differences being so great in external characters and relatively so slight in internal parts or organs. Man can hardly select, or only with much difficulty, any deviation of structure excepting such as is externally visible; and indeed he rarely cares for what is internal. He can never act by selection, excepting on variations which are first given to him in some slight degree by nature. No man would ever try to make a fantail, till he saw a pigeon with a tail developed in some slight degree in an unusual manner, or a pouter till he saw a pigeon with a crop of somewhat unusual size; and the more abnormal or unusual any character was when it first appeared, the more likely it would be to catch his attention. But to use such an expression as trying to make a fantail, is, I have no doubt, in most cases, utterly incorrect. The man who first selected a pigeon with a slightly larger tail, never dreamed what the descendants of that pigeon would become through long-continued, partly unconscious and partly methodical selection. Perhaps the parent bird of all fantails had only fourteen tail-feathers somewhat expanded, like the present Java fantail, or like individuals of other and distinct breeds, in which as many as seventeen tail-feathers have been counted. Perhaps the first pouter-pigeon did not inflate its crop much more than the turbit now does the upper part of its œsophagus,—a habit which is disregarded by all fanciers, as it is not one of the points of the breed.

Nor let it be thought that some great deviation of structure would be necessary to catch the fancier's eye: he perceives extremely small differences, and it is in human nature to value any novelty, however slight, in one's own possession. Nor must the value which would formerly be set on any slight differences in the individuals of the same species, be judged of by the value which would now be set on them, after several breeds have once fairly been established. Many slight differences might, and indeed do now, arise amongst pigeons, which are rejected as faults or deviations from the standard of perfection of each breed. The common goose has not given rise to any marked varieties; hence the Thoulouse and the common breed, which differ only in colour, that most fleeting of characters, have lately been exhibited as distinct at our poultry-shows.

I think these views further explain what has sometimes been noticed—namely, that we know nothing about the origin or history of any of our domestic breeds. But, in fact, a breed, like a dialect of a language, can hardly be said to have had a definite origin. A man preserves and breeds from an individual with some slight deviation of structure, or takes more care than usual in matching his best animals and thus improves them, and the improved individuals slowly spread in the immediate neighbourhood. But as yet they will hardly have a distinct name, and from being only slightly valued, their history will be disregarded. When further improved by the same slow and gradual process, they will spread more widely, and will get recognised as something distinct and valuable, and will then probably first receive a provincial name. In semi-civilised countries, with little free communication, the spreading and knowledge of any new sub-breed will be a slow process. As soon as the points of value of the new sub-breed are once fully acknowledged, the principle, as I have called it, of unconscious selection will always tend,—perhaps more at one period than at another, as the breed rises or falls in fashion,—perhaps more in one district than in another, according to the state of civilisation of the inhabitants,—slowly to add to the characteristic features of the breed, whatever they may be. But the chance will be infinitely small of any record having been preserved of such slow, varying, and insensible changes.

I must now say a few words on the circumstances, favourable, or the reverse, to man's power of selection. A high degree of variability is obviously favourable, as freely giving the materials for selection to work on; not that mere individual differences are not amply sufficient, with extreme care, to allow of the accumulation of a large amount of modification in almost any desired direction. But as variations manifestly useful or pleasing to man appear only occasionally, the chance of their appearance will be much increased by a large number of individuals being kept; and hence this comes to be of the highest importance to success. On this principle Marshall has remarked, with respect to the sheep of parts of Yorkshire, that " as they generally belong to poor people, and are mostly *in small lots,* they never can be improved." On the other hand, nurserymen, from raising large stocks of the same plants, are generally far more successful than amateurs in getting new and valuable varieties. The keeping of a large number of individuals of a species in any country requires that the species should be placed under favourable conditions of life, so as to breed freely in that country. When the individuals of any species are scanty, all the individuals, whatever their quality may be, will generally be allowed to breed, and this will effectually prevent selection. But probably the most important point of all, is, that the animal or plant should be so highly useful to man, or so much valued by him, that the closest attention should be paid to even the slightest deviation in the qualities or structure of each individual. Unless such attention be paid nothing can be effected. I have seen it gravely remarked, that it was most fortunate that the strawberry began to vary just when gardeners began to attend closely to this plant. No doubt the strawberry had always varied since it was cultivated, but the slight varieties had been neglected. As soon, however, as gardeners picked out individual plants with slightly larger, earlier, or better fruit, and raised seedlings from them, and again picked out the best seedlings and bred from them, then, there appeared (aided by some crossing with distinct species) those many admirable varieties of the strawberry which have been raised during the last thirty or forty years.

In the case of animals with separate sexes, facility in preventing crosses is an important element of success in the formation of new races,—at least, in a country which is already stocked with other races. In this respect enclosure of the land plays a part. Wandering savages or the inhabitants of open plains rarely possess more than one breed of the same species. Pigeons can be mated for life, and this is a great convenience to the fancier, for thus many races may be kept true, though mingled in the same aviary; and this circumstance must have largely favoured the improvement and formation of new breeds. Pigeons, I may add, can be propagated in great numbers and at a very quick rate, and inferior birds may be freely rejected, as when killed they serve for food. On the other hand, cats, from their nocturnal rambling habits, cannot be matched, and, although so much valued by women and children, we hardly ever see a distinct breed kept up; such breeds as we do sometimes see are almost always imported from some other country, often from islands. Although I do not doubt that some domestic animals vary less than others, yet the rarity or absence of distinct breeds of the cat, the donkey, peacock, goose, &c., may be attributed in main part to selection not having been brought into play: in cats, from the difficulty in pairing them; in donkeys, from only a few being kept by poor people, and little attention paid to their breeding; in peacocks, from not being very easily reared and a large stock not kept; in geese, from being valuable only for two purposes, food and feathers, and more especially from no pleasure having been felt in the display of distinct breeds.

To sum up on the origin of our Domestic Races of animals and plants. I believe that the conditions of life, from their action on the reproductive system, are so far of the highest importance as causing variability. I do not believe that variability is an inherent and necessary contingency, under all circumstances, with all organic beings, as some authors have thought. The effects of variability are modified by various degrees of inheritance and of reversion. Variability is governed by many unknown laws, more especially by that of correlation of growth. Something may be attributed to the direct action of the conditions of life. Something must be attributed to use and disuse. The final result is thus rendered infinitely complex. In some cases, I do not doubt that the intercrossing of species, aboriginally distinct, has played an important part in the origin of our domestic productions. When in any country several domestic breeds have once been established, their occasional intercrossing, with the aid of selection, has, no doubt, largely aided in the formation of new sub-breeds; but the importance of the crossing of varieties has, I believe, been greatly exaggerated, both in regard to animals and to those plants which are propagated by seed. In plants which are temporarily propagated by cuttings, buds, &c., the importance of the crossing both of distinct species and of varieties is immense; for the cultivator here quite disregards the extreme variability both of hybrids and mongrels, and the frequent sterility of hybrids; but the cases of plants not propagated by seed are of little importance to us, for their endurance is only temporary. Over all these causes of Change I am convinced that the accumulative action of Selection, whether applied methodically and more quickly, or unconsciously and more slowly, but more efficiently, is by far the predominant Power.

15

Reprinted from pages 192–220 and 233–249 of *The Variation of Animals and Plants under Domestication,* vol. II, John Murray, London, 1868

SELECTION BY MAN

C. R. Darwin

SELECTION A DIFFICULT ART — METHODICAL, UNCONSCIOUS, AND NATURAL SELECTION — RESULTS OF METHODICAL SELECTION — CARE TAKEN IN SELECTION — SELECTION WITH PLANTS — SELECTION CARRIED ON BY THE ANCIENTS, AND BY SEMI-CIVILISED PEOPLE — UNIMPORTANT CHARACTERS OFTEN ATTENDED TO — UNCONSCIOUS SELECTION — AS CIRCUMSTANCES SLOWLY CHANGE, SO HAVE OUR DOMESTICATED ANIMALS CHANGED THROUGH THE ACTION OF UNCONSCIOUS SELECTION — INFLUENCE OF DIFFERENT BREEDERS ON THE SAME SUB-VARIETY — PLANTS AS AFFECTED BY UNCONSCIOUS SELECTION — EFFECTS OF SELECTION AS SHOWN BY THE GREAT AMOUNT OF DIFFERENCE IN THE PARTS MOST VALUED BY MAN.

THE power of Selection, whether exercised by man, or brought into play under nature through the struggle for existence and the consequent survival of the fittest, absolutely depends on the variability of organic beings. Without variability nothing can be effected; slight individual differences, however, suffice for the work, and are probably the sole differences which are effective in the production of new species. Hence our discussion on the causes and laws of variability ought in strict order to have preceded our present subject, as well as the previous subjects of inheritance, crossing, &c.; but practically the present arrangement has been found the most convenient. Man does not attempt to cause variability; though he unintentionally effects this by exposing organisms to new conditions of life, and by crossing breeds already formed. But variability being granted, he works wonders. Unless some degree of selection be exercised, the free commingling of the individuals of the same variety soon obliterates, as we have previously seen, the slight differences which may arise, and gives to the whole body of individuals uniformity of character. In separated districts, long-continued exposure to different conditions of life may perhaps produce new races without the aid of selection; but to this difficult subject of the direct action of the conditions of life we shall in a future chapter recur.

When animals or plants are born with some conspicuous and firmly inherited new character, selection is reduced to the preservation of such individuals, and to the subsequent prevention of crosses; so that nothing more need be said on the subject. But in the great majority of cases a new character, or some superiority in an old character, is at first faintly pronounced, and is not strongly inherited; and then the full difficulty of selection is experienced. Indomitable patience, the finest powers of discrimination, and sound judgment must be exercised during many years. A clearly predetermined object must be kept steadily in view. Few men are endowed with all these qualities, especially with that of discriminating very slight differences; judgment can be acquired only by long experience; but if any of these qualities be wanting, the labour of a life may be thrown away. I have been astonished when celebrated breeders, whose skill and judgment have been proved by their success at exhibitions, have shown me their animals, which appeared all alike, and have assigned their reasons for matching this and that individual. The importance of the great principle of Selection mainly lies in this power of selecting scarcely appreciable differences, which nevertheless are found to be transmissible, and which can be accumulated until the result is made manifest to the eyes of every beholder.

The principle of selection may be conveniently divided into three kinds. *Methodical selection* is that which guides a man who systematically endeavours to modify a breed according to some predetermined standard. *Unconscious selection* is that which follows from men naturally preserving the most valued and destroying the less valued individuals, without any thought of altering the breed; and undoubtedly this process slowly works great changes. Unconscious selection graduates into methodical, and only extreme cases can be distinctly separated; for he who preserves a useful or perfect animal will generally breed from it with the hope of getting offspring of the same character; but as long as he has not a predetermined purpose to improve the breed, he may be said to be selecting unconsciously.[1] Lastly, we have *Natural selection*, which implies that the individuals which are best fitted for the complex, and in the course of ages changing conditions to which they are exposed, generally survive and procreate their kind. With domestic productions, with which alone we are here strictly concerned, natural selection comes to a certain extent into action, independently of, and even in opposition to, the will of man.

Methodical Selection.—What man has effected within recent times in England by methodical selection is clearly shown by our exhibitions of improved quadrupeds and fancy birds. With respect to cattle, sheep, and pigs, we owe their great improvement to a long series of well-known names—Bakewell, Colling, Ellman, Bates, Jonas Webb, Lords Leicester and Western, Fisher Hobbs, and others. Agricultural writers are unanimous on the power of selection: any number of statements to this effect could be quoted; a few will suffice. Youatt, a sagacious and experienced observer, writes,[2] the principle of selection is "that which enables the agriculturist, not only to modify the character of his flock, but to change it altogether." A great breeder of shorthorns[3] says, "In the anatomy of the shoulder "modern breeders have made great improvements on the "Ketton shorthorns by correcting the defect in the knuckle or "shoulder-joint, and by laying the top of the shoulder more "snugly into the crop, and thereby filling up the hollow "behind it. . . . . The eye has its fashion at different periods: "at one time the eye high and outstanding from the head, and "at another time the sleepy eye sunk into the head; but these "extremes have merged into the medium of a full, clear, and "prominent eye with a placid look."

Again, hear what an excellent judge of pigs[4] says: "The legs

[1] The term *unconscious selection* has been objected to as a contradiction: but *see* some excellent observations on this head by Prof. Huxley ('Nat. Hist. Review,' Oct. 1864, p. 578), who remarks that when the wind heaps up sand-dunes it sifts and *unconsciously selects* from the gravel on the beach grains of sand of equal size.

[2] Sheep, 1838, p. 60.

[3] Mr. J. Wright on Shorthorn Cattle, in 'Journal of Royal Agricult. Soc.,' vol. vii. pp. 208, 209.

[4] H. D. Richardson on Pigs, 1847, p. 44.

" should be no longer than just to prevent the animal's belly " from trailing on the ground. The leg is the least profitable " portion of the hog, and we therefore require no more of it than " is absolutely necessary for the support of the rest." Let any one compare the wild-boar with any improved breed, and he will see how effectually the legs have been shortened.

Few persons, except breeders, are aware of the systematic care taken in selecting animals, and of the necessity of having a clear and almost prophetic vision into futurity. Lord Spencer's skill and judgment were well known; and he writes,[5] " It is " therefore very desirable, before any man commences to breed " either cattle or sheep, that he should make up his mind to the " shape and qualities he wishes to obtain, and steadily pursue " this object." Lord Somerville, in speaking of the marvellous improvement of the New Leicester sheep, effected by Bakewell and his successors, says, " It would seem as if they had first drawn a perfect form, and then given it life." Youatt[6] urges the necessity of annually drafting each flock, as many animals will certainly degenerate " from the standard of excellence, which the breeder has established in his own mind." Even with a bird of such little importance as the canary, long ago (1780-1790) rules were established, and a standard of perfection was fixed, according to which the London fanciers tried to breed the several sub-varieties.[7] A great winner of prizes at the Pigeon-shows,[8] in describing the Short-faced Almond Tumbler, says, " There are many first-rate fanciers who are particularly partial " to what is called the goldfinch-beak, which is very beautiful; " others say, take a full-size round cherry, then take a barley-" corn, and judiciously placing and thrusting it into the cherry, " form as it were your beak; and that is not all, for it will form " a good head and beak, provided, as I said before, it is judi-" ciously done; others take an oat; but as I think the gold-" finch-beak the handsomest, I would advise the inexperienced " fancier to get the head of a goldfinch, and keep it by him " for his observation." Wonderfully different as is the beak of the rock-pigeon and goldfinch, undoubtedly, as far as ex-

[5] 'Journal of R. Agricult. Soc.,' vol. i. p. 24.
[6] Sheep, pp. 520, 319.
[7] Loudon's 'Mag. of Nat. Hist.,' vol. viii., 1835, p. 618.
[8] 'A Treatise on the Art of Breeding the Almond Tumbler,' 1851, p. 9.

ternal shape and proportions are concerned, the end has been nearly gained.

Not only should our animals be examined with the greatest care whilst alive, but, as Anderson remarks,[9] their carcases should be scrutinised, "so as to breed from the descendants of such only as, in the language of the butcher, cut up well." The "grain of the meat" in cattle, and its being well marbled with fat,[10] and the greater or less accumulation of fat in the abdomen of our sheep, have been attended to with success. So with poultry, a writer,[11] speaking of Cochin-China fowls, which are said to differ much in the quality of their flesh, says, "the best "mode is to purchase two young brother-cocks, kill, dress, and "serve up one; if he be indifferent, similarly dispose of the "other, and try again; if, however, he be fine and well-flavoured, "his brother will not be amiss for breeding purposes for the "table."

The great principle of the division of labour has been brought to bear on selection. In certain districts[12] "the breeding of "bulls is confined to a very limited number of persons, who by "devoting their whole attention to this department, are able "from year to year to furnish a class of bulls which are steadily "improving the general breed of the district." The rearing and letting of choice rams has long been, as is well known, a chief source of profit to several eminent breeders. In parts of Germany this principle is carried with merino sheep to an extreme point.[13] "So important is the proper selection of "breeding animals considered, that the best flock-masters do "not trust to their own judgment, or to that of their shepherds, "but employ persons called 'sheep-classifiers,' who make it their "special business to attend to this part of the management of "several flocks, and thus to preserve, or if possible to improve, "the best qualities of both parents in the lambs." In Saxony, "when the lambs are weaned, each in his turn is placed upon "a table that his wool and form may be minutely observed.

[9] 'Recreations in Agriculture,' vol. ii. p. 409.

[10] Youatt on Cattle, pp. 191, 227.

[11] Ferguson, 'Prize Poultry,' 1854, p. 208.

[12] Wilson, in 'Transact. Highland Agricult. Soc.,' quoted in 'Gard. Chronicle,' 1844, p. 29.

[13] Simmonds, quoted in 'Gard. Chronicle,' 1855, p. 637. And for the second quotation, *see* Youatt on Sheep, p. 171.

"The finest are selected for breeding and receive a first mark. "When they are one year old, and prior to shearing them, "another close examination of those previously marked takes "place: those in which no defect can be found receive a second "mark, and the rest are condemned. A few months afterwards "a third and last scrutiny is made; the prime rams and ewes "receive a third and final mark, but the slightest blemish is "sufficient to cause the rejection of the animal." These sheep are bred and valued almost exclusively for the fineness of their wool; and the result corresponds with the labour bestowed on their selection. Instruments have been invented to measure accurately the thickness of the fibres; and "an Austrian fleece has been produced of which twelve hairs equalled in thickness one from a Leicester sheep."

Throughout the world, wherever silk is produced, the greatest care is bestowed on selecting the cocoons from which the moths for breeding are to be reared. A careful cultivator[14] likewise examines the moths themselves, and destroys those that are not perfect. But what more immediately concerns us is that certain families in France devote themselves to raising eggs for sale.[15] In China, near Shanghai, the inhabitants of two small districts have the privilege of raising eggs for the whole surrounding country, and that they may give up their whole time to this business, they are interdicted by law from producing silk.[16]

The care which successful breeders take in matching their birds is surprising. Sir John Sebright, whose fame is perpetuated by the "Sebright Bantam," used to spend "two and three days in examining, consulting, and disputing with a friend which were the best of five or six birds."[17] Mr. Bult, whose pouter-pigeons won so many prizes and were exported to North America under the charge of a man sent on purpose, told me that he always deliberated for several days before he matched each pair. Hence we can understand the advice of an eminent fancier, who writes,[18] "I would here particularly guard

[14] Robinet, 'Vers à Soie,' 1848, p. 271.

[15] Quatrefages, 'Les Maladies du Ver à Soie,' 1859, p. 101.

[16] M. Simon, in 'Bull. de la Soc. d'Acclimat.,' tom. ix., 1862, p. 221.

[17] 'The Poultry Chronicle,' vol. i., 1854, p. 607.

[18] J. M. Eaton, 'A Treatise on Fancy Pigeons,' 1852, p. xiv., and 'A Treatise on the Almond Tumbler,' 1851, p. 11.

"you against having too great a variety of pigeons, otherwise "you will know a little of all, but nothing about one as it "ought to be known." Apparently it transcends the power of the human intellect to breed all kinds: "it is possible that "there may be a few fanciers that have a good general know- "ledge of fancy pigeons; but there are many more who labour "under the delusion of supposing they know what they do not." The excellence of one sub-variety, the Almond Tumbler, lies in the plumage, carriage, head, beak, and eye; but it is too presumptuous in the beginner to try for all these points. The great judge above quoted says, "there are some young fanciers "who are over-covetous, who go for all the above five properties "at once; they have their reward by getting nothing." We thus see that breeding even fancy pigeons is no simple art: we may smile at the solemnity of these precepts, but he who laughs will win no prizes.

What methodical selection has effected for our animals is sufficiently proved, as already remarked, by our Exhibitions. So greatly were the sheep belonging to some of the earlier breeders, such as Bakewell and Lord Western, changed, that many persons could not be persuaded that they had not been crossed. Our pigs, as Mr. Corringham remarks,[19] during the last twenty years have undergone, through rigorous selection together with crossing, a complete metamorphosis. The first exhibition for poultry was held in the Zoological Gardens in 1845; and the improvement effected since that time has been great. As Mr. Baily, the great judge, remarked to me, it was formerly ordered that the comb of the Spanish cock should be upright, and in four or five years all good birds had upright combs; it was ordered that the Polish cock should have no comb or wattles, and now a bird thus furnished would be at once disqualified; beards were ordered, and out of fifty-seven pens lately (1860) exhibited at the Crystal Palace, all had beards. So it has been in many other cases. But in all cases the judges order only what is occasionally produced and what can be improved and rendered constant by selection. The steady increase of weight during the last few years in our

[19] 'Journal Royal Agricultural Soc.,' vol. vi. p. 22.

fowls, turkeys, ducks, and geese is notorious; "six-pound ducks are now common, whereas four pounds was formerly the average." As the actual time required to make a change has not often been recorded, it may be worth mentioning that it took Mr. Wicking thirteen years to put a clean white head on an almond tumbler's body, "a triumph," says another fancier, "of which he may be justly proud." [20]

Mr. Tollet, of Betley Hall, selected cows, and especially bulls, descended from good milkers, for the sole purpose of improving his cattle for the production of cheese; he steadily tested the milk with the lactometer, and in eight years he increased, as I was informed by him, the product in the proportion of four to three. Here is a curious case [21] of steady but slow progress, with the end not as yet fully attained: in 1784 a race of silk-worms was introduced into France, in which one hundred out of the thousand failed to produce white cocoons; but now, after careful selection during sixty-five generations, the proportion of yellow cocoons has been reduced to thirty-five in the thousand.

With plants selection has been followed with the same good results as with animals. But the process is simpler, for plants in the great majority of cases bear both sexes. Nevertheless, with most kinds it is necessary to take as much care to prevent crosses as with animals or unisexual plants; but with some plants, such as peas, this care does not seem to be necessary. With all improved plants, excepting of course those which are propagated by buds, cuttings, &c., it is almost indispensable to examine the seedlings and destroy those which depart from the proper type. This is called "roguing," and is, in fact, a form of selection, like the rejection of inferior animals. Experienced horticulturists and agriculturists incessantly urge every one to preserve the finest plants for the production of seed.

Although plants often present much more conspicuous variations than animals, yet the closest attention is generally requisite to detect each slight and favourable change. Mr. Masters relates [22] how "many a patient hour was devoted," whilst he was

[20] 'Poultry Chronicle,' vol. ii., 1855, p. 596.
[21] Isid. Geoffroy St. Hilaire, 'Hist. Nat. Gén.,' tom. iii. p. 254.
[22] 'Gardener's Chronicle,' 1850, p. 198.

young, to the detection of differences in peas intended for seed. Mr. Barnet[23] remarks that the old scarlet American strawberry was cultivated for more than a century without producing a single variety; and another writer observes how singular it was that when gardeners first began to attend to this fruit it began to vary; the truth no doubt being that it had always varied, but that, until slight varieties were selected and propagated by seed, no conspicuous result was obtained. The finest shades of difference in wheat have been discriminated and selected with almost as much care, as we see in Colonel Le Couteur's works, as in the case of the higher animals; but with our cereals the process of selection has seldom or never been long continued.

It may be worth while to give a few examples of methodical selection with plants; but in fact the great improvement of all our anciently cultivated plants may be attributed to selection long carried on, in part methodically, and in part unconsciously. I have shown in a former chapter how the weight of the gooseberry has been increased by systematic selection and culture. The flowers of the Heartsease have been similarly increased in size and regularity of outline. With the Cineraria, Mr. Glenny[24] "was bold enough, when the flowers were ragged and starry "and ill defined in colour, to fix a standard which was then "considered outrageously high and impossible, and which, even "if reached, it was said, we should be no gainers by, as it would "spoil the beauty of the flowers. He maintained that he was "right; and the event has proved it to be so." The doubling of flowers has several times been effected by careful selection: the Rev. W. Williamson,[25] after sowing during several years seed of *Anemone coronaria*, found a plant with one additional petal; he sowed the seed of this, and by perseverance in the same course obtained several varieties with six or seven rows of petals. The single Scotch rose was doubled, and yielded eight good varieties in nine or ten years.[26] The Canterbury bell (*Campanula medium*) was doubled by careful selection in four generations.[27] In four years Mr. Buckman,[28] by culture and

[23] 'Transact. Hort. Soc.,' vol. vi. p. 152.

[24] 'Journal of Horticulture,' 1862, p. 369.

[25] 'Transact. Hort. Soc.,' vol. iv. p. 381.

[26] 'Transact. Hort. Soc.,' vol. iv. p. 285.

[27] Rev. W. Bromehead, in 'Gard. Chronicle,' 1857, p. 550.

[28] 'Gard. Chronicle,' 1862, p. 721.

careful selection, converted parsnips, raised from wild seed, into a new and good variety. By selection during a long course of years, the early maturity of peas has been hastened from ten to twenty-one days.[29] A more curious case is offered by the beet-plant, which, since its cultivation in France, has almost exactly doubled its yield of sugar. This has been effected by the most careful selection; the specific gravity of the roots being regularly tested, and the best roots saved for the production of seed.[30]

Selection by Ancient and Semi-civilised People.

In attributing so much importance to the selection of animals and plants, it may be objected that methodical selection would not have been carried on during ancient times. A distinguished naturalist considers it as absurd to suppose that semi-civilised people should have practised selection of any kind. Undoubtedly the principle has been systematically acknowledged and followed to a far greater extent within the last hundred years than at any former period, and a corresponding result has been gained; but it would be a great error to suppose, as we shall immediately see, that its importance was not recognised and acted on during the most ancient times, and by semi-civilised people. I should premise that many facts now to be given only show that care was taken in breeding; but when this is the case, selection is almost sure to be practised to a certain extent. We shall hereafter be enabled better to judge how far selection, when only occasionally carried on, by a few of the inhabitants of a country, will slowly produce a great effect.

In a well-known passage in the thirtieth chapter of Genesis, rules are given for influencing, as was then thought possible, the colour of sheep; and speckled and dark breeds are spoken of as being kept separate. By the time of David the fleece was likened to snow. Youatt,[31] who has discussed all the passages in relation to breeding in the Old Testament, concludes that

[29] Dr. Anderson, in 'The Bee,' vol. vi. p. 96; Mr. Barnes, in 'Gard. Chronicle,' 1844, p. 476.

[30] Godron, 'De l'Espèce,' 1859, tom. ii. p. 69; 'Gard. Chronicle,' 1854, p. 258.

[31] On Sheep, p. 18.

at this early period "some of the best principles of breeding must have been steadily and long pursued." It was ordered, according to Moses, that "Thou shalt not let thy cattle gender with a diverse kind;" but mules were purchased,[32] so that at this early period other nations must have crossed the horse and ass. It is said[33] that Erichthonius, some generations before the Trojan war, had many brood-mares, "which by his care and judgment in the choice of stallions produced a breed of horses superior to any in the surrounding countries." Homer (Book v.) speaks of Æneas's horses as bred from mares which were put to the steeds of Laomedon. Plato, in his 'Republic,' says to Glaucus, "I see that you raise at your house a great many dogs for the chase. Do you take care about breeding and pairing them? Among animals of good blood, are there not always some which are superior to the rest?" To which Glaucus answers in the affirmative.[34] Alexander the Great selected the finest Indian cattle to send to Macedonia to improve the breed.[35] According to Pliny,[36] King Pyrrhus had an especially valuable breed of oxen; and he did not suffer the bulls and cows to come together till four years old, that the breed might not degenerate. Virgil, in his Georgics (lib. iii.), gives as strong advice as any modern agriculturist could do, carefully to select the breeding stock; "to note the tribe, the lineage, and the sire; whom to reserve for husband of the herd;"—to brand the progeny;—to select sheep of the purest white, and to examine if their tongues are swarthy. We have seen that the Romans kept pedigrees of their pigeons, and this would have been a senseless proceeding had not great care been taken in breeding them. Columella gives detailed instructions about breeding fowls: "Let the breeding hens therefore be of a choice colour, "a robust body, square-built, full-breasted, with large heads, "with upright and bright-red combs. Those are believed to be "the best bred which have five toes."[37] According to Tacitus, the Celts attended to the races of their domestic animals;

[32] Volz, 'Beiträge zur Kulturgeschichte,' 1852, s. 47.

[33] Mitford's 'History of Greece,' vol. i. p. 73.

[34] Dr. Dally, translated in 'Anthropological Review,' May 1864, p. 101.

[35] Volz, 'Beiträge,' &c., 1852, s. 80.

[36] 'History of the World,' ch. 45.

[37] 'Gardener's Chronicle,' 1848, p. 323.

and Cæsar states that they paid high prices to merchants for fine imported horses.[38] In regard to plants, Virgil speaks of yearly culling the largest seeds; and Celsus says, "where the corn and crop is but small, we must pick out the best ears of corn, and of them lay up our seed separately by itself." [39]

Coming down the stream of time, we may be brief. At about the beginning of the ninth century Charlemagne expressly ordered his officers to take great care of his stallions; and if any proved bad or old, to forewarn him in good time before they were put to the mares.[40] Even in a country so little civilised as Ireland during the ninth century, it would appear from some ancient verses,[41] describing a ransom demanded by Cormac, that animals from particular places, or having a particular character, were valued. Thus it is said,—

> Two pigs of the pigs of Mac Lir,
> A ram and ewe both round and red,
> I brought with me from Aengus.
> I brought with me a stallion and a mare
> From the beautiful stud of Manannan,
> A bull and a white cow from Druim Cain.

Athelstan, in 930, received as a present from Germany, running-horses; and he prohibited the exportation of English horses. King John imported "one hundred chosen stallions from Flanders." [42] On June 16th, 1305, the Prince of Wales wrote to the Archbishop of Canterbury, begging for the loan of any choice stallion, and promising its return at the end of the season.[43] There are numerous records at ancient periods in English history of the importation of choice animals of various kinds, and of foolish laws against their exportation. In the reigns of Henry VII. and VIII. it was ordered that the magistrates, at Michaelmas, should scour the heaths and commons, and destroy all mares beneath a certain size.[44] Some of our earlier kings passed laws against the slaughtering rams of any good breed before they were seven years old, so that they

[38] Reynier, 'De l'Economie des Celtes,' 1818, pp. 487, 503.

[39] Le Couteur on Wheat, p. 15.

[40] Michel, 'Des Haras,' 1861, p. 84.

[41] Sir W. Wilde, an 'Essay on Unmanufactured Animal Remains,' &c., 1860, p. 11.

[42] Col. Hamilton Smith, 'Nat. Library,' vol. xii., Horses, pp. 135, 140.

[43] Michel, 'Des Haras,' p. 90.

[44] Mr. Baker, 'History of the Horse,' Veterinary, vol. xiii. p. 423.

might have time to breed. In Spain Cardinal Ximenes issued, in 1509, regulations on the *selection* of good rams for breeding.[45]

The Emperor Akbar Khan before the year 1600 is said to have "wonderfully improved" his pigeons by crossing the breeds; and this necessarily implies careful selection. About the same period the Dutch attended with the greatest care to the breeding of these birds. Belon in 1555 says that good managers in France examined the colour of their goslings in order to get geese of a white colour and better kinds. Markham in 1631 tells the breeder "to elect the largest and goodliest conies," and enters into minute details. Even with respect to seeds of plants for the flower-garden, Sir J. Hanmer writing about the year 1660[46] says, in "choosing seed, the best seed is the most weighty, and is had from the lustiest and most vigorous stems;" and he then gives rules about leaving only a few flowers on plants for seed; so that even such details were attended to in our flower-gardens two hundred years ago. In order to show that selection has been silently carried on in places where it would not have been expected, I may add that in the middle of the last century, in a remote part of North America, Mr. Cooper improved by careful selection all his vegetables, "so that they were greatly superior to those of any "other person. When his radishes, for instance, are fit for use, "he takes ten or twelve that he most approves, and plants "them at least 100 yards from others that blossom at the same "time. In the same manner he treats all his other plants, "varying the circumstances according to their nature."[47]

In the great work on China published in the last century by the Jesuits, and which is chiefly compiled from ancient Chinese encyclopædias, it is said that with sheep "improving the breed "consists in choosing with particular care the lambs which are "destined for propagation, in nourishing them well, and in "keeping the flocks separate." The same principles were applied by the Chinese to various plants and fruit-trees.[48] An

[45] M. l'Abbé Carlier, in 'Journal de Physique,' vol. xxiv., 1784, p. 181: this memoir contains much information on the ancient selection of sheep; and is my authority for rams not being killed young in England.

[46] 'Gardener's Chronicle,' 1843, p. 389.

[47] Communications to Board of Agriculture, quoted in Dr. Darwin's 'Phytologia,' 1800, p. 451.

[48] 'Mémoire sur les Chinois,' 1786, tom. xi. p. 55; tom. v. p. 507.

imperial edict recommends the choice of seed of remarkable size; and selection was practised even by imperial hands, for it is said that the Ya-mi, or imperial rice, was noticed at an ancient period in a field by the Emperor Khang-hi, was saved and cultivated in his garden, and has since become valuable from being the only kind which will grow north of the Great Wall.[49] Even with flowers, the tree pæony (*P. moutan*) has been cultivated, according to Chinese traditions, for 1400 years; between 200 and 300 varieties have been raised, which are cherished like tulips formerly were by the Dutch.[50]

Turning now to semi-civilised people and to savages: it occurred to me, from what I had seen of several parts of South America, where fences do not exist, and where the animals are of little value, that there would be absolutely no care in breeding or selecting them; and this to a large extent is true. Roulin,[51] however, describes in Colombia a naked race of cattle, which are not allowed to increase, on account of their delicate constitution. According to Azara[52] horses are often born in Paraguay with curly hair; but, as the natives do not like them, they are destroyed. On the other hand, Azara states that a hornless bull, born in 1770, was preserved and propagated its race. I was informed of the existence in Banda Oriental of a breed with reversed hair; and the extraordinary niata cattle first appeared and have since been kept distinct in La Plata. Hence certain conspicuous variations have been preserved, and others have been habitually destroyed, in these countries, which are so little favourable for careful selection. We have also seen that the inhabitants sometimes introduce cattle on their estates to prevent the evil effects of close interbreeding. On the other hand, I have heard on reliable authority that the Gauchos of the Pampas never take any pains in selecting the best bulls or stallions for breeding; and this probably accounts for the cattle and horses being remarkably uniform in character throughout the immense range of the Argentine republic.

Looking to the Old World, in the Sahara Desert "The "Touareg is as careful in the selection of his breeding Mahari

[49] 'Recherches sur l'Agriculture des Chinois,' par L. D'Hervey-Saint-Denys, 1850, p. 229. With respect to Khang-hi, see Huc's 'Chinese Empire,' p. 311.

[50] Anderson, in 'Linn. Transact.,' vol. xii. p. 253.

[51] 'Mém. de l'Acad.' (divers savans), tom. vi., 1835, p. 333.

[52] 'Des Quadrupèdes du Paraguay,' 1801, tom. ii. p. 333, 371.

"(a fine race of the dromedary) as the Arab is in that of his "horse. The pedigrees are handed down, and many a dromedary "can boast a genealogy far longer than the descendants of the "Darley Arabian."[53] According to Pallas the Mongolians endeavour to breed the Yaks or horse-tailed buffaloes with white tails, for these are sold to the Chinese mandarins as fly-flappers; and Moorcroft, about seventy years after Pallas, found that white-tailed animals were still selected for breeding.[54]

We have seen in the chapter on the Dog that savages in different parts of North America and in Guiana cross their dogs with wild Canidæ, as did the ancient Gauls, according to Pliny. This was done to give their dogs strength and vigour, in the same way as the keepers in large warrens now sometimes cross their ferrets (as I have been informed by Mr. Yarrell) with the wild polecat, "to give them more devil." According to Varro, the wild ass was formerly caught and crossed with the tame animal to improve the breed, in the same manner as at the present day the natives of Java sometimes drive their cattle into the forests to cross with the wild Banteng (*Bos sondaicus*).[55] In Northern Siberia, among the Ostyaks the dogs vary in markings in different districts, but in each place they are spotted black and white in a remarkably uniform manner;[56] and from this fact alone we may infer careful breeding, more especially as the dogs of one locality are famed throughout the country for their superiority. I have heard of certain tribes of Esquimaux who take pride in their teams of dogs being uniformly coloured. In Guiana, as Sir R. Schomburgk informs me,[57] the dogs of the Turuma Indians are highly valued and extensively bartered: the price of a good one is the same as that given for a wife: they are kept in a sort of cage, and the Indians "take great care when the female is in season to prevent her uniting with a dog of an inferior description." The Indians told Sir Robert that, if a dog proved bad or useless,

[53] 'The Great Sahara,' by the Rev. H. B. Tristram, 1860, p. 238.

[54] Pallas, 'Act. Acad. St. Petersburg,' 1777, p. 249; Moorcroft and Trebeck, 'Travels in the Himalayan Provinces,' 1841.

[55] Quoted from Raffles, in the 'Indian Field,' 1859, p. 196: for Varro, *see* Pallas, *ut supra*.

[56] Erman's 'Travels in Siberia,' Eng. translat., vol. i. p. 453.

[57] *See* also 'Journal of R. Geograph. Soc.,' vol. xiii. part i. p. 65.

he was not killed, but was left to die from sheer neglect. Hardly any nation is more barbarous than the Fuegians, but I hear from Mr. Bridges, the Catechist to the Mission, that, "when these savages have a large, strong, and active bitch, they "take care to put her to a fine dog, and even take care to feed "her well, that her young may be strong and well favoured."

In the interior of Africa, negroes, who have not associated with white men, show great anxiety to improve their animals: they "always choose the larger and stronger males for stock:" the Malakolo were much pleased at Livingstone's promise to send them a bull, and some Bakalolo carried a live cock all the way from Loanda into the interior.[58] Further south on the same continent, Andersson states that he has known a Damara give two fine oxen for a dog which struck his fancy. The Damaras take great delight in having whole droves of cattle of the same colour, and they prize their oxen in proportion to the size of their horns. "The Namaquas have a perfect mania for "a uniform team; and almost all the people of Southern Africa "value their cattle next to their women, and take a pride in "possessing animals that look high-bred." "They rarely or "never make use of a handsome animal as a beast of burden."[59] The power of discrimination which these savages possess is wonderful, and they can recognise to which tribe any cattle belong. Mr. Andersson further informs me that the natives frequently match a particular bull with a particular cow.

The most curious case of selection by semi-civilised people, or indeed by any people, which I have found recorded, is that given by Garcilazo de la Vega, a descendant of the Incas, as having been practised in Peru before the country was subjugated by the Spaniards.[60] The Incas annually held great hunts, when all the wild animals were driven from an immense circuit to a central point. The beasts of prey were first destroyed as injurious. The wild Guanacos and Vicunas were sheared; the old males and females killed, and the others set at liberty. The various kinds of deer were examined; the old males and females

[58] Livingstone's 'First Travels,' pp. 191, 439, 565: *see* also 'Expedition to the Zambesi,' 1865, p. 495, for an analogous case respecting a good breed of goats.

[59] Andersson's 'Travels in South Africa,' pp. 232, 318, 319.

[60] Dr. Vavasseur, in 'Bull. de la Soc. d'Acclimat.,' tom. viii., 1861, p. 136.

were likewise killed; "but the young females, with a certain number of males, selected from the most beautiful and strong," were given their freedom. Here, then, we have selection by man aiding natural selection. So that the Incas followed exactly the reverse system of that which our Scottish sportsmen are accused of following, namely, of steadily killing the finest stags, thus causing the whole race to degenerate.[61] In regard to the domesticated llamas and alpacas, they were separated in the time of the Incas according to colour; and if by chance one in a flock was born of the wrong colour, it was eventually put into another flock.

In the genus Auchenia there are four forms,—the Guanaco and Vicuna, found wild and undoubtedly distinct species; the Llama and Alpaca, known only in a domesticated condition. These four animals appear so different, that most professed naturalists, especially those who have studied these animals in their native country, maintain that they are specifically distinct, notwithstanding that no one pretends to have seen a wild llama or alpaca. Mr. Ledger, however, who has closely studied these animals both in Peru and during their exportation to Australia, and who has made many experiments on their propagation, adduces arguments[62] which seem to me conclusive, that the llama is the domesticated descendant of the guanaco, and the alpaca of the vicuna. And now that we know that these animals many centuries ago were systematically bred and selected, there is nothing surprising in the great amount of change which they have undergone.

It appeared to me at one time probable that, though ancient and semi-civilised people might have attended to the improvement of their more useful animals in essential points, yet that they would have disregarded unimportant characters. But human nature is the same throughout the world: fashion everywhere reigns supreme, and man is apt to value whatever he may chance to possess. We have seen that in South America the niata cattle, which certainly are not made useful by their shortened faces and upturned nostrils, have been preserved. The Damaras of South Africa value their cattle for uniformity

[61] 'The Natural History of Dee Side,' 1855, p. 476.
[62] 'Bull. de la Soc. d'Acclimat.,' tom. vii., 1860, p. 457.

of colour and enormously long horns. The Mongolians value their yaks for their white tails. And I shall now show that there is hardly any peculiarity in our most useful animals which, from fashion, superstition, or some other motive, has not been valued, and consequently preserved. With respect to cattle, "an early record," according to Youatt,[63] speaks of a hundred "white cows with red ears being demanded as a compensation "by the princes of North and South Wales. If the cattle were "of a dark or black colour, 150 were to be presented." So that colour was attended to in Wales before its subjugation by England. In Central Africa, an ox that beats the ground with its tail is killed; and in South Africa some of the Damaras will not eat the flesh of a spotted ox. The Kaffirs value an animal with a musical voice; and "at a sale in British Kaffraria the "low of a heifer excited so much admiration that a sharp com-"petition sprung up for her possession, and she realised a "considerable price."[64] With respect to sheep, the Chinese prefer rams without horns; the Tartars prefer them with spirally wound horns, because the hornless are thought to lose courage.[65] Some of the Damaras will not eat the flesh of hornless sheep. In regard to horses, at the end of the fifteenth century animals of the colour described as *liart pommé* were most valued in France. The Arabs have a proverb, "Never buy a horse with four white feet, for he carries his shroud with him;"[66] the Arabs also, as we have seen, despise dun-coloured horses. So with dogs, Xenophon and others at an ancient period were prejudiced in favour of certain colours; and "white or slate-coloured hunting dogs were not esteemed."[67]

Turning to poultry, the old Roman gourmands thought that the liver of a white goose was the most savoury. In Paraguay black-skinned fowls are kept because they are thought to be more productive, and their flesh the most proper for invalids.[68] In Guiana, as I am informed by Sir R. Schomburgk, the aborigines will not eat the flesh or eggs of the fowl, but two

63 'Cattle,' p. 48.

64 Livingstone's Travels, p. 576; Andersson, 'Lake Ngami,' 1856, p. 222. With respect to the sale in Kaffraria, see 'Quarterly Review,' 1860, p. 139.

65 'Mémoire sur les Chinois' (by the Jesuits), 1786, tom. xi. p. 57.

66 F. Michel, 'Des Haras,' pp. 47, 50.

67 Col. Hamilton Smith, Dogs, in 'Nat. Lib.,' vol. x. p. 103.

68 Azara, 'Quadrupèdes du Paraguay,' tom. ii. p. 324.

races are kept distinct merely for ornament. In the Philippines, no less than nine sub-varieties of the game cock are kept and named, so that they must be separately bred.

At the present time in Europe, the smallest peculiarities are carefully attended to in our most useful animals, either from fashion, or as a mark of purity of blood. Many examples could be given, two will suffice. "In the Western counties of England "the prejudice against a white pig is nearly as strong as against "a black one in Yorkshire." In one of the Berkshire sub-breeds, it is said, "the white should be confined to four white feet, "a white spot between the eyes, and a few white hairs behind "each shoulder." Mr. Saddler possessed "three hundred pigs, "every one of which was marked in this manner."[69] Marshall, towards the close of the last century, in speaking of a change in one of the Yorkshire breeds of cattle, says the horns have been considerably modified, as "a clean, small, sharp horn has been *fashionable* for the last twenty years."[70] In a part of Germany the cattle of the Race de Gfoehl are valued for many good qualities, but they must have horns of a particular curvature and tint, so much so that mechanical means are applied if they take a wrong direction; but the inhabitants "consider it "of the highest importance that the nostrils of the bull should "be flesh-coloured, and the eyelashes light; this is an indis-"pensable condition. A calf with blue nostrils would not be "purchased, or purchased at a very low price."[71] Therefore let no man say that any point or character is too trifling to be methodically attended to and selected by breeders.

Unconscious Selection.—By this term I mean, as already more than once explained, the preservation by man of the most valued, and the destruction of the least valued individuals, without any conscious intention on his part of altering the breed. It is difficult to offer direct proofs of the results which follow from this kind of selection; but the indirect evidence is abundant. In fact, except that in the one case man acts intentionally, and in the other unintentionally, there is little difference between

[69] Sidney's edit. of Youatt, 1860, pp. 24, 25.
[70] 'Rural Economy of Yorkshire vol. ii. p. 182.
[71] Moll et Gayot, 'Du Bœuf,' 1860, p. 547.

methodical and unconscious selection. In both cases man preserves the animals which are most useful or pleasing to him, and destroys or neglects the others. But no doubt a far more rapid result follows from methodical than from unconscious selection. The "roguing" of plants by gardeners, and the destruction by law in Henry VIII.'s reign of all under-sized mares, are instances of a process the reverse of selection in the ordinary sense of the word, but leading to the same general result. The influence of the destruction of individuals having a particular character is well shown by the necessity of killing every lamb with a trace of black about it, in order to keep the flock white; or again, by the effects on the average height of the men of France of the destructive wars of Napoleon, by which many tall men were killed, the short ones being left to be the fathers of families. This at least is the conclusion of those who have closely studied the subject of the conscription; and it is certain that since Napoleon's time the standard for the army has been lowered two or three times.

Unconscious selection so blends into methodical that it is scarcely possible to separate them. When a fancier long ago first happened to notice a pigeon with an unusually short beak, or one with the tail-feathers unusually developed, although he bred from these birds with the distinct intention of propagating the variety, yet he could not have intended to make a short-faced tumbler or a fantail, and was far from knowing that he had made the first step towards this end. If he could have seen the final result, he would have been struck with astonishment, but, from what we know of the habits of fanciers, probably not with admiration. Our English carriers, barbs, and short-faced tumblers have been greatly modified in the same manner, as we may infer both from the historical evidence given in the chapters on the Pigeon, and from the comparison of birds brought from distant countries.

So it has been with dogs; our present fox-hounds differ from the old English hound; our greyhounds have become lighter; the wolf-dog, which belonged to the greyhound class, has become extinct; the Scotch deer-hound has been modified, and is now rare. Our bulldogs differ from those which were formerly used for baiting bulls. Our pointers and Newfoundlands do not closely resemble any native dog now found in the countries whence they were brought. These changes have been effected partly by crosses; but in every case the result has been governed by the strictest selection. Nevertheless there is no reason to suppose that man intentionally and methodically made the breeds exactly what they now are. As our horses became fleeter, and the country more cultivated and smoother, fleeter fox-hounds were desired and produced, but probably without any one distinctly foreseeing what they would become. Our pointers and setters, the latter almost certainly descended from large spaniels, have been greatly modified in accordance with fashion and the desire for increased speed. Wolves have become extinct, deer have become rarer, bulls are no longer baited, and the corresponding breeds of the dog have answered to the change. But we may feel almost sure that when, for instance, bulls were no longer baited, no man said to himself, I will now breed my dogs of smaller size, and thus create the present race. As circumstances changed, men unconsciously and slowly modified their course of selection.

With race-horses selection for swiftness has been followed methodically, and our horses can now easily beat their progenitors. The increased size and different appearance of the English race-horse led a good observer in India to ask, "Could any one in this year of 1856, looking at our race-horses, conceive that they were the result of the union of the Arab horse and the African mare?"[72] This change has, it is probable, been largely effected through unconscious selection, that is, by the general wish to breed as fine horses as possible in each generation, combined with training and high feeding, but without any intention to give to them their present appearance. According to Youatt,[73] the introduction in Oliver Cromwell's time of three celebrated Eastern stallions speedily affected the English breed; "so that Lord Harleigh, one of the old school, complained that the great horse was fast disappearing." This is an excellent proof how carefully selection must have been attended to; for without such care, all traces of so small an infusion of Eastern blood would soon have been absorbed and

[72] 'The India Sporting Review,' vol. ii. p. 181; 'The Stud Farm,' by Cecil, p. 58.
[73] 'The Horse,' p. 22.

lost. Notwithstanding that the climate of England has never been esteemed particularly favourable to the horse, yet long-continued selection, both methodical and unconscious, together with that practised by the Arabs during a still longer and earlier period, has ended in giving us the best breed of horses in the world. Macaulay[74] remarks, "Two men whose authority on such "subjects was held in great esteem, the Duke of Newcastle and "Sir John Fenwick, pronounced that the meanest hack ever "imported from Tangier would produce a finer progeny than "could be expected from the best sire of our native breed. "They would not readily have believed that a time would "come when the princes and nobles of neighbouring lands "would be as eager to obtain horses from England as ever the "English had been to obtain horses from Barbary."

The London dray-horse, which differs so much in appearance from any natural species, and which from its size has so astonished many Eastern princes, was probably formed by the heaviest and most powerful animals having been selected during many generations in Flanders and England, but without the least intention or expectation of creating a horse such as we now see. If we go back to an early period of history, we behold in the antique Greek statues, as Schaaffhausen has remarked,[75] a horse equally unlike a race or dray horse, and differing from any existing breed.

The results of unconscious selection, in an early stage, are well shown in the difference between the flocks descended from the same stock, but separately reared by careful breeders. Youatt gives an excellent instance of this fact in the sheep belonging to Messrs. Buckley and Burgess, which "have been "purely bred from the original stock of Mr. Bakewell for "upwards of fifty years. There is not a suspicion existing in "the mind of any one at all acquainted with the subject that "the owner of either flock has deviated in any one instance "from the pure blood of Mr. Bakewell's flock; yet the differ-"ence between the sheep possessed by these two gentlemen "is so great, that they have the appearance of being quite "different varieties."[76] I have seen several analogous and well-

[74] 'History of England,' vol. i. p. 316.

[75] 'Ueber Beständigkeit der Arten.'

[76] Youatt on Sheep, p. 315.

marked cases with pigeons: for instance, I had a family of barbs, descended from those long bred by Sir J. Sebright, and another family long bred by another fancier, and the two families plainly differed from each other. Nathusius—and a more competent witness could not be cited—observes that, though the Shorthorns are remarkably uniform in appearance (except in colouring), yet that the individual character and wishes of each breeder become impressed on his cattle, so that different herds differ slightly from each other.[77] The Hereford cattle assumed their present well-marked character soon after the year 1769, through careful selection by Mr. Tomkins,[78] and the breed has lately split into two strains—one strain having a white face, and differing slightly, it is said,[79] in some other points; but there is no reason to believe that this split, the origin of which is unknown, was intentionally made; it may with much more probability be attributed to different breeders having attended to different points. So again, the Berkshire breed of swine in the year 1810 had greatly changed from what it had been in 1780; and since 1810 at least two distinct sub-breeds have borne this same name.[80] When we bear in mind how rapidly all animals increase, and that some must be annually slaughtered and some saved for breeding, then, if the same breeder during a long course of years deliberately settles which shall be saved and which shall be killed, it is almost inevitable that his individual frame of mind will influence the character of his stock, without his having had any intention to modify the breed or form a new strain.

Unconscious selection in the strictest sense of the word, that is, the saving of the more useful animals and the neglect or slaughter of the less useful, without any thought of the future, must have gone on occasionally from the remotest period and amongst the most barbarous nations. Savages often suffer from famines, and are sometimes expelled by war from their own homes. In such cases it can hardly be doubted that they would save their most useful animals. When the Fuegians

77 'Ueber Shorthorn Rindvieh,' 1857, s. 51.

78 Low, 'Domesticated Animals,' 1845, p. 363.

79 'Quarterly Review,' 1849, p. 392.

80 H. von Nathusius, 'Vorstudien . . . . Schweineschædel,' 1864, s. 140.

are hard pressed by want, they kill their old women for food rather than their dogs; for, as we were assured, "old women no use—dogs catch otters." The same sound sense would surely lead them to preserve their more useful dogs when still harder pressed by famine. Mr. Oldfield, who has seen so much of the aborigines of Australia, informs me that "they are all very glad to get a European kangaroo dog, and several instances have been known of the father killing his own infant that the mother might suckle the much-prized puppy." Different kinds of dogs would be useful to the Australian for hunting opossums and kangaroos, and to the Fuegian for catching fish and otters; and the occasional preservation in the two countries of the most useful animals would ultimately lead to the formation of two widely distinct breeds.

With plants, from the earliest dawn of civilisation, the best variety which at each period was known would generally have been cultivated and its seeds occasionally sown; so that there will have been some selection from an extremely remote period, but without any prefixed standard of excellence or thought of the future. We at the present day profit by a course of selection occasionally and unconsciously carried on during thousands of years. This is proved in an interesting manner by Oswald Heer's researches on the lake-inhabitants of Switzerland, as given in a former chapter; for he shows that the grain and seed of our present varieties of wheat, barley, oats, peas, beans, lentils, and poppy, exceed in size those which were cultivated in Switzerland during the Neolithic and Bronze periods. These ancient people, during the Neolithic period, possessed also a crab considerably larger than that now growing wild on the Jura.[81] The pears described by Pliny were evidently extremely inferior in quality to our present pears. We can realise the effects of long-continued selection and cultivation in another way, for would any one in his senses expect to raise a first-rate apple from the seed of a truly wild crab, or a luscious melting pear from the wild pear? Alphonse De Candolle informs me that he has lately seen on an ancient mosaic at Rome a representation of

[81] *See* also Dr. Christ, in 'Rütimeyer's Pfahlbauten,' 1861, s. 226.

the melon; and as the Romans, who were such gourmands, are silent on this fruit, he infers that the melon has been greatly ameliorated since the classical period.

Coming to later times, Buffon,[82] on comparing the flowers, fruit, and vegetables which were then cultivated, with some excellent drawings made a hundred and fifty years previously, was struck with surprise at the great improvement which had been effected; and remarks that these ancient flowers and vegetables would now be rejected, not only by a florist but by a village gardener. Since the time of Buffon the work of improvement has steadily and rapidly gone on. Every florist who compares our present flowers with those figured in books published not long since, is astonished at the change. A well-known amateur,[83] in speaking of the varieties of Pelargonium raised by Mr. Garth only twenty-two years before, remarks, "what a rage they excited: surely we had attained perfection, "it was said; and now not one of the flowers of those days "will be looked at. But none the less is the debt of gratitude "which we owe to those who saw what was to be done, and did "it." Mr. Paul, the well-known horticulturist, in writing of the same flower,[84] says he remembers when young being delighted with the portraits in Sweet's work; "but what are they in point "of beauty compared with the Pelargoniums of this day? Here "again nature did not advance by leaps; the improvement "was gradual, and, if we had neglected those very gradual "advances, we must have foregone the present grand results." How well this practical horticulturist appreciates and illustrates the gradual and accumulative force of selection! The Dahlia has advanced in beauty in a like manner; the line of improvement being guided by fashion, and by the successive modifications which the flower slowly underwent.[85] A steady and gradual change has been noticed in many other flowers: thus an old florist,[86] after describing the leading varieties of the Pink which were grown in 1813, adds, "the pinks of those days would now "be scarcely grown as border-flowers." The improvement of

[82] The passage is given 'Bull. Soc. d'Acclimat.,' 1858, p. 11.

[83] 'Journal of Horticulture,' 1862, p. 394.

[84] 'Gardener's Chronicle,' 1857, p. 85.

[85] *See* Mr. Wildman's address to the Floricult. Soc., in 'Gardener's Chronicle,' 1843, p. 86.

[86] 'Journal of Horticulture,' Oct. 24th, 1865, p. 239.

so many flowers and the number of the varieties which have been raised is all the more striking when we hear that the earliest known flower-garden in Europe, namely at Padua, dates only from the year 1545.[87]

Effects of Selection, as shown by the parts most valued by man presenting the greatest amount of Difference.—The power of long-continued selection, whether methodical or unconscious, or both combined, is well shown in a general way, namely, by the comparison of the differences between the varieties of distinct species, which are valued for different parts, such as for the leaves, or stems, or tubers, the seed, or fruit, or flowers. Whatever part man values most, that part will be found to present the greatest amount of difference. With trees cultivated for their fruit, Sageret remarks that the fruit is larger than in the parent-species, whilst with those cultivated for the seed, as with nuts, walnuts, almonds, chesnuts, &c., it is the seed itself which is larger; and he accounts for this fact by the fruit in the one case, and by the seed in the other, having been carefully attended to and selected during many ages. Gallesio has made the same observation. Godron insists on the diversity of the tuber in the potato, of the bulb in the onion, and of the fruit in the melon; and on the close similarity in these same plants of the other parts.[88]

In order to judge how far my own impression on this subject was correct, I cultivated numerous varieties of the same species close to each other. The comparison of the amount of difference between widely different organs is necessarily vague; I will therefore give the results in only a few cases. We have previously seen in the ninth chapter how greatly the varieties of the cabbage differ in their foliage and stems, which are the selected parts, and how closely they resemble each other in their flowers, capsules, and seeds. In seven varieties of the radish, the roots differed greatly in colour and shape, but no difference

[87] Prescott's 'Hist. of Mexico,' vol. ii. p. 61.

[88] Sageret, 'Pomologie Physiologique,' 1830, p. 47; Gallesio, 'Teoria della Riproduzione,' 1816, p. 88; Godron, 'De l'Espèce,' 1859, tom. ii. pp. 63, 67, 70. In my tenth and eleventh chapters I have given details on the potato; and I can confirm similar remarks with respect to the onion. I have also shown how far Naudin concurs in regard to the varieties of the melon.

whatever could be detected in their foliage, flowers, or seeds. Now what a contrast is presented, if we compare the flowers of the varieties of these two plants with those of any species cultivated in our flower-gardens for ornament; or if we compare their seeds with those of the varieties of maize, peas, beans, &c., which are valued and cultivated for their seeds. In the ninth chapter it was shown that the varieties of the pea differ but little except in the tallness of the plant, moderately in the shape of the pod, and greatly in the pea itself, and these are all selected points. The varieties, however, of the *Pois sans parchemin* differ much more in their pods, and these are eaten and valued. I cultivated twelve varieties of the common bean; one alone, the Dwarf Fan, differed considerably in general appearance; two differed in the colour of their flowers, one being an albino, and the other being wholly instead of partially purple; several differed considerably in the shape and size of the pod, but far more in the bean itself, and this is the valued and selected part. Toker's bean, for instance, is twice-and-a-half as long and broad as the horse-bean, and is much thinner and of a different shape.

The varieties of the gooseberry, as formerly described, differ much in their fruit, but hardly perceptibly in their flowers or organs of vegetation. With the plum, the differences likewise appear to be greater in the fruit than in the flowers or leaves. On the other hand, the seed of the strawberry, which corresponds with the fruit of the plum, differs hardly at all; whilst every one knows how greatly the fruit—that is, the enlarged receptacle—differs in the several varieties. In apples, pears, and peaches the flowers and leaves differ considerably, but not, as far as I can judge, in proportion with the fruit. The Chinese double-flowering peaches, on the other hand, show that varieties of this tree have been formed, which differ more in the flower than in fruit. If, as is highly probable, the peach is the modified descendant of the almond, a surprising amount of change has been effected in the same species, in the fleshy covering of the former and in the kernels of the latter.

When parts stand in such close relation to each other as the fleshy covering of the fruit (whatever its homological nature may be) and the seed, when one part is modified, so generally is the other, but by no means necessarily in the same degree. With the plum-tree, for instance, some varieties produce plums which are nearly alike, but include stones extremely dissimilar in shape; whilst conversely other varieties produce dissimilar fruit with barely distinguishable stones; and generally the stones, though they have never been subjected to selection, differ greatly in the several varieties of the plum. In other cases organs which are not manifestly related, through some unknown bond vary together, and are consequently liable, without any intention on man's part, to be simultaneously acted on by selection. Thus the varieties of the stock (Matthiola) have been selected solely for the beauty of their flowers, but the seeds differ greatly in colour and somewhat in size. Varieties of the lettuce have been selected solely on account of their leaves, yet produce seeds which likewise differ in colour. Generally, through the law of correlation, when a variety differs greatly from its fellow-varieties in any one character, it differs to a certain extent in several other characters. I observed this fact when I cultivated together many varieties of the same species, for I used first to make a list of the varieties which differed most from each other in their foliage and manner of growth, afterwards of those that differed most in their flowers, then in their seed-capsules, and lastly in their mature seed; and I found that the same names generally occurred in two, three, or four of the successive lists. Nevertheless the greatest amount of difference between the varieties was always exhibited, as far as I could judge, by that part or organ for which the plant was cultivated.

When we bear in mind that each plant was at first cultivated because useful to man, and that its variation was a subsequent, often a long subsequent, event, we cannot explain the greater amount of diversity in the valuable parts by supposing that species endowed with an especial tendency to vary in any particular manner, were originally chosen. We must attribute the result to the variations in these parts having been successively preserved, and thus continually augmented; whilst other variations, excepting such as inevitably appeared through correlation, were neglected and lost. Hence we may infer that most plants might be made, through long-continued selection, to yield races as different from each other in any character as they now are in those parts for which they are valued and cultivated.

With animals we see something of the same kind; but they have not been domesticated in sufficient number or yielded sufficient varieties for a fair comparison. Sheep are valued for their wool, and the wool differs much more in the several races than the hair in cattle. Neither sheep, goats, European cattle, nor pigs are valued for their fleetness or strength; and we do not possess breeds differing in these respects like the race-horse and dray-horse. But fleetness and strength are valued in camels and dogs; and we have with the former the swift dromedary and heavy camel; with the latter the greyhound and mastiff. But dogs are valued even in a higher degree for their mental qualities and senses; and every one knows how greatly the races differ in these respects. On the other hand, where the dog is valued solely to serve for food, as in the Polynesian islands and China, it is described as an extremely stupid animal.[89] Blumenbach remarks that "many dogs, such as the badger-" dog, have a build so marked and so appropriate for particular "purposes, that I should find it very difficult to persuade myself "that this astonishing figure was an accidental consequence of "degeneration."[90] But had Blumenbach reflected on the great principle of selection, he would not have used the term degeneration, and he would not have been astonished that dogs and other animals should become excellently adapted for the service of man.

On the whole we may conclude that whatever part or character is most valued—whether the leaves, stems, tubers, bulbs, flowers, fruit, or seed of plants, or the size, strength, fleetness, hairy covering, or intellect of animals—that character will almost invariably be found to present the greatest amount of difference both in kind and degree. And this result may be safely attributed to man having preserved during a long course of generations the variations which were useful to him, and neglected the others.

[89] Godron, 'De l'Espèce,' tom. ii. p. 27.
[90] 'The Anthropological Treatises of Blumenbach,' 1865, p. 292.

[*Editor's Note:* Material has been omitted at this point.]

Circumstances favourable to Selection by Man.

The possibility of selection rests on variability, and this, as we shall see in the following chapters, mainly depends on changed conditions of life, but is governed by infinitely complex, and, to a great extent, unknown laws. Domestication, even when long continued, occasionally causes but a small amount of variability, as in the case of the goose and turkey. The slight differences, however, which characterise each individual animal and plant would in most, probably in all cases, suffice for the production of distinct races through careful and prolonged selection. We see what selection, though acting on mere individual differences, can effect when families of cattle, sheep, pigeons, &c., of the same race, have been separately bred during a number of years by different men without any wish on their part to modify the breed. We see the same fact in the difference between hounds bred for hunting in different districts,[39] and in many other such cases.

In order that selection should produce any result, it is manifest that the crossing of distinct races must be prevented; hence facility in pairing, as with the pigeon, is highly favourable for the work; and difficulty in pairing, as with cats, prevents the formation of distinct breeds. On nearly the same principle the cattle of the small island of Jersey have been improved in their milking qualities "with a rapidity that could not have been obtained in a widely extended country like France."[40] Although free crossing is a danger on the one side which every one can see, too close interbreeding is a hidden danger on the other side. Unfavourable conditions of life overrule the power of selection. Our improved heavy breeds of cattle and sheep could not have been formed on mountainous pastures; nor could dray-horses have been raised on a barren and inhospitable land, such as the Falkland islands, where even the light horses of La Plata rapidly decrease in size. Nor could the wool of sheep have been much increased in length within the Tropics; yet selection has kept Merino sheep nearly true under diversified and unfavourable conditions of life. The power of selection is so great, that breeds of the dog, sheep, and poultry, of the largest and least size, long and short beaked pigeons, and other breeds with opposite characters, have had their characteristic qualities augmented, though treated in every way alike, being exposed to the same climate and fed on the same food. Selection, however, is either checked or favoured by the effects of use or habit. Our wonderfully-improved pigs could never have been formed if they had been forced to search for their own food; the English racehorse and greyhound could not have been improved up to their present high standard of excellence without constant training.

As conspicuous deviations of structure occur rarely, the improvement of each breed is generally the result, as already

[39] 'Encyclop. of Rural Sports,' p. 405.
[40] Col. Le Couteur, 'Journal Roy. Agricult. Soc.,' vol. iv. p. 43.

remarked, of the selection of slight individual differences. Hence the closest attention, the sharpest powers of observation, and indomitable perseverance, are indispensable. It is, also, highly important that many individuals of the breed which is to be improved should be raised; for thus there will be a better chance of the appearance of variations in the right direction, and individuals varying in an unfavourable manner may be freely rejected or destroyed. But that a large number of individuals should be raised, it is necessary that the conditions of life should favour the propagation of the species. Had the peacock been bred as easily as the fowl, we should probably ere this have had many distinct races. We see the importance of a large number of plants, from the fact of nursery gardeners almost always beating amateurs in the exhibition of new varieties. In 1845 it was estimated[41] that between 4000 and 5000 pelargoniums were annually raised from seed in England, yet a decidedly improved variety is rarely obtained. At Messrs. Carter's grounds, in Essex, where such flowers as the Lobelia, Nemophila, Mignonette, &c., are grown by the acre for seed, "scarcely a season passes without some new kinds being raised, or some improvement effected on old kinds.[42] At Kew, as Mr. Beaton remarks, where many seedlings of common plants are raised, "you see new forms of Laburnums, Spiræas, and other shrubs."[43] So with animals: Marshall,[44] in speaking of the sheep in one part of Yorkshire, remarks, "as they belong to poor people, and are mostly in small lots, they never can be improved." Lord Rivers, when asked how he succeeded in always having first-rate greyhounds, answered, "I breed many, and hang many." This, as another man remarks, "was the secret of his success; and the same will be found in exhibiting fowls, —successful competitors breed largely, and keep the best."[45]

It follows from this that the capacity of breeding at an early age and at short successive intervals, as with pigeons, rabbits, &c., facilitates selection; for the result is thus soon made visible, and perseverance in the work is encouraged. It can hardly be

[41] 'Gardener's Chronicle,' 1845, p. 273.

[42] 'Journal of Horticulture,' 1862, p. 157.

[43] 'Cottage Gardener,' 1860, p. 368.

[44] 'A Review of Reports,' 1808, p. 406.

[45] 'Gardener's Chronicle,' 1853, p. 45.

accidental that the great majority of the culinary and agricultural plants which have yielded numerous races are annuals or biennials, which therefore are capable of rapid propagation and thus of improvement. Sea-kale, asparagus, common and Jerusalem artichokes, potatoes, and onions, alone are perennials. Onions are propagated like annuals, and of the other plants just specified, none, with the exception of the potato, have yielded more than one or two varieties. No doubt fruit-trees, which cannot be propagated quickly by seed, have yielded a host of varieties, though not permanent races; but these, judging from pre-historic remains, were produced at a later and more civilised epoch than the races of culinary and agricultural plants.

A species may be highly variable, but distinct races will not be formed, if from any cause selection be not applied. The carp is highly variable, but it would be extremely difficult to select slight variations in fishes whilst living in their natural state, and distinct races have not been formed;[46] on the other hand, a closely allied species, the gold-fish, from being reared in glass or open vessels, and from having been carefully attended to by the Chinese, has yielded many races. Neither the bee, which has been semi-domesticated from an extremely remote period, nor the cochineal insect, which was cultivated by the aboriginal Mexicans, has yielded races; and it would be impossible to match the queen-bee with any particular drone, and most difficult to match cochineal insects. Silk-moths, on the other hand, have been subjected to rigorous selection, and have produced a host of races. Cats, which from their nocturnal habits cannot be selected for breeding, do not, as formerly remarked, yield distinct races in the same country. The ass in England varies much in colour and size; but it is an animal of little value, bred by poor people; consequently there has been no selection, and distinct races have not been formed. We must not attribute the inferiority of our asses to climate, for in India they are of even smaller size than in Europe. But when selection is brought to bear on the ass, all is changed. Near Cordova, as I am informed (Feb. 1860) by Mr. W. E. Webb, C.E., they are carefully bred, as much as 200*l.* having been paid for a stallion ass,

[46] Isidore Geoffroy St. Hilaire, 'Hist. Nat. Gén.,' tom. iii. p. 49. On the Cochineal Insect, p. 46.

and they have been immensely improved. In Kentucky, asses have been imported (for breeding mules) from Spain, Malta, and France; these "seldom averaged more than fourteen hands "high; but the Kentuckians, by great care, have raised them "up to fifteen hands, and sometimes even to sixteen. The prices "paid for these splendid animals, for such they really are, will "prove how much they are in request. One male, of great "celebrity, was sold for upwards of one thousand pounds sterling." These choice asses are sent to cattle-shows, one day being given to their exhibition.[47]

Analogous facts have been observed with plants: the nutmeg-tree in the Malay archipelago is highly variable, but there has been no selection, and there are no distinct races.[48] The common mignonette (*Reseda odorata*), from bearing inconspicuous flowers, valued solely for their fragrance, "remains in the same unim-"proved condition as when first introduced."[49] Our common forest-trees are very variable, as may be seen in every extensive nursery-ground; but as they are not valued like fruit-trees, and as they seed late in life, no selection has been applied to them; consequently, as Mr. Patrick Matthews remarks,[50] they have not yielded distinct races, leafing at different periods, growing to different sizes, and producing timber fit for different purposes. We have gained only some fanciful and semi-monstrous varieties, which no doubt appeared suddenly as we now see them.

Some botanists have argued that plants cannot have so strong a tendency to vary as is generally supposed, because many species long grown in botanic gardens, or unintentionally cultivated year after year mingled with our corn crops, have not produced distinct races; but this is accounted for by slight variations not having been selected and propagated. Let a plant which is now grown in a botanic garden, or any common weed, be cultivated on a large scale, and let a sharp-sighted gardener look out for each slight variety and sow the seed, and then, if distinct races are not produced, the argument will be valid.

[47] Capt. Marryat, quoted by Blyth in 'Journ. Asiatic Soc. of Bengal,' vol. xxviii. p. 229.

[48] Mr. Oxley, 'Journal of the Indian Archipelago,' vol. ii., 1848, p. 645.

[49] Mr. Abbey, in 'Journal of Horticulture,' Dec. 1, 1863, p. 430.

[50] 'On Naval Timber,' 1831, p. 107.

The importance of selection is likewise shown by considering special characters. For instance, with most breeds of fowls the form of the comb and the colour of the plumage have been attended to, and are eminently characteristic of each race; but in Dorkings, fashion has never demanded uniformity of comb or colour; and the utmost diversity in these respects prevails. Rose-combs, double-combs, cup-combs, &c., and colours of all kinds, may be seen in purely-bred and closely related Dorking fowls, whilst other points, such as the general form of body, and the presence of an additional toe, have been attended to, and are invariably present. It has also been ascertained that colour can be fixed in this breed, as well as in any other.[51]

During the formation or improvement of a breed, its members will always be found to vary much in those characters to which especial attention is directed, and of which each slight improvement is eagerly sought and selected. Thus with short-faced tumbler-pigeons, the shortness of the beak, shape of head and plumage,—with carriers, the length of the beak and wattle,—with fantails, the tail and carriage,—with Spanish fowls, the white face and comb,—with long-eared rabbits, the length of ear, are all points which are eminently variable. So it is in every case, and the large price paid for first-rate animals proves the difficulty of breeding them up to the highest standard of excellence. This subject has been discussed by fanciers,[52] and the greater prizes given for highly improved breeds, in comparison with those given for old breeds which are not now undergoing rapid improvement, has been fully justified. Nathusius makes[53] a similar remark when discussing the less uniform character of improved Shorthorn cattle and of the English horse, in comparison, for example, with the unennobled cattle of Hungary, or with the horses of the Asiatic steppes. This want of uniformity in the parts which at the time are undergoing selection, chiefly depends on the strength of the principle of reversion; but it likewise depends to a certain extent on the continued

[51] Mr. Baily, in 'The Poultry Chronicle,' vol. ii., 1854, p. 150. Also vol. i. p. 342; vol. iii. p. 245.

[52] 'Cottage Gardener,' 1855, December, p. 171; 1856, January, pp. 248, 323.

[53] 'Ueber Shorthorn Rindvieh,' 1857, s. 51.

variability of the parts which have recently varied. That the same parts do continue varying in the same manner we must admit, for, if it were not so, there could be no improvement beyond an early standard of excellence, and we know that such improvement is not only possible, but is of general occurrence.

As a consequence of continued variability, and more especially of reversion, all highly improved races, if neglected or not subjected to incessant selection, soon degenerate. Youatt gives a curious instance of this in some cattle formerly kept in Glamorganshire; but in this case the cattle were not fed with sufficient care. Mr. Baker, in his memoir on the Horse, sums up: "It must have "been observed in the preceding pages that, whenever there has "been neglect, the breed has proportionally deteriorated."[54] If a considerable number of improved cattle, sheep, or other animals of the same race, were allowed to breed freely together, with no selection, but with no change in their condition of life, there can be no doubt that after a score or hundred generations they would be very far from excellent of their kind; but, from what we see of the many common races of dogs, cattle, fowls, pigeons, &c., which without any particular care have long retained nearly the same character, we have no grounds for believing that they would altogether depart from their type.

It is a general belief amongst breeders that characters of all kinds become fixed by long-continued inheritance. But I have attempted to show in the fourteenth chapter that this belief apparently resolves itself into the following proposition, namely, that all characters whatever, whether recently acquired or ancient, tend to be transmitted, but that those which have already long withstood all counteracting influences, will, as a general rule, continue to withstand them, and consequently be faithfully transmitted.

Tendency in Man to carry the practice of Selection to an extreme point.

It is an important principle that in the process of selection man almost invariably wishes to go to an extreme point. Thus, in useful qualities, there is no limit to his desire to breed certain

[54] 'The Veterinary,' vol. xiii. p. 720. For the Glamorganshire cattle, *see* Youatt on Cattle, p. 51.

horses and dogs as fleet as possible, and others as strong as possible; certain kinds of sheep for extreme fineness, and others for extreme length of wool; and he wishes to produce fruit, grain, tubers, and other useful parts of plants, as large and excellent as possible. With animals bred for amusement, the same principle is even more powerful; for fashion, as we see even in our dress, always runs to extremes. This view has been expressly admitted by fanciers. Instances were given in the chapters on the pigeon, but here is another: Mr. Eaton, after describing a comparatively new variety, namely, the Archangel, remarks, "What "fanciers intend doing with this bird I am at a loss to know, "whether they intend to breed it down to the tumbler's head "and beak, or carry it out to the carrier's head and beak; leaving "it as they found it, is not progressing." Ferguson, speaking of fowls, says, "their peculiarities, whatever they may be, must "necessarily be fully developed: a little peculiarity forms nought "but ugliness, seeing it violates the existing laws of symmetry." So Mr. Brent, in discussing the merits of the sub-varieties of the Belgian canary-bird, remarks, "Fanciers always go to extremes; "they do not admire indefinite properties."[55]

This principle, which necessarily leads to divergence of character, explains the present state of various domestic races. We can thus see how it is that race-horses and dray-horses, greyhounds and mastiffs, which are opposed to each other in every character,—how varieties so distinct as Cochin-China fowls and bantams, or carrier-pigeons with very long beaks, and tumblers with excessively short beaks, have been derived from the same stock. As each breed is slowly improved, the inferior varieties are first neglected and finally lost. In a few cases, by the aid of old records, or from intermediate varieties still existing in countries where other fashions have prevailed, we are enabled partially to trace the graduated changes through which certain breeds have passed. Selection, whether methodical or unconscious, always tending towards an extreme point, together with the neglect and slow extinction of the intermediate and less-valued forms, is the key which unlocks the mystery how man has produced such wonderful results.

[55] J. M. Eaton, 'A Treatise on Fancy Pigeons,' p. 82; Ferguson, on 'Rare and Prize Poultry,' p. 162; Mr. Brent, in 'Cottage Gardener,' Oct. 1860, p. 13.

In a few instances selection, guided by utility for a single purpose, has led to convergence of character. All the improved and different races of the pig, as Nathusius has well shown,[56] closely approach each other in character, in their shortened legs and muzzles, their almost hairless, large, rounded bodies, and small tusks. We see some degree of convergence in the similar outline of the body in well-bred cattle belonging to distinct races.[57] I know of no other such cases.

Continued divergence of character depends on, and is indeed a clear proof, as previously remarked, of the same parts continuing to vary in the same direction. The tendency to mere general variability or plasticity of organisation can certainly be inherited, even from one parent, as has been shown by Gärtner and Kölreuter, in the production of varying hybrids from two species, of which one alone was variable. It is in itself probable that, when an organ has varied in any manner, it will again vary in the same manner, if the conditions which first caused the being to vary remain, as far as can be judged, the same. This is either tacitly or expressly admitted by all horticulturists: if a gardener observes one or two additional petals in a flower, he feels confident that in a few generations he will be able to raise a double flower, crowded with petals. Some of the seedlings from the weeping Moccas oak were so prostrate that they only crawled along the ground. A seedling from the fastigate or upright Irish yew is described as differing greatly from the parent-form "by the exaggeration of the fastigate habit of its branches."[58] Mr. Sheriff, who has been more successful than any other man in raising new kinds of wheat, remarks, "A good variety may safely be regarded as the forerunner of a better one."[59] A great rose-grower, Mr. Rivers, has made the same remark with respect to roses. Sageret,[60] who had large experience, in speaking of the future progress of fruit-trees, observes that the most important principle is "that the more plants have departed from their original type, the more they tend to depart from it." There is apparently much truth in this

[56] 'Die Racen des Schweines,' 1860, s. 48.

[57] *See* some good remarks on this head by M. de Quatrefages, 'Unité de l'Espèce Humaine,' 1861, p. 119.

[58] Verlot, 'Des Variétés,' 1865, p. 94.

[59] Mr. Patrick Sheriff, in 'Gard. Chronicle,' 1858, p. 771.

[60] 'Pomologie Physiolog.,' 1830, p. 106.

remark; for we can in no other way understand the surprising amount of difference between varieties in the parts or qualities which are valued, whilst other parts retain nearly their original character.

The foregoing discussion naturally leads to the question, what is the limit to the possible amount of variation in any part or quality, and, consequently, is there any limit to what selection can effect? Will a race-horse ever be reared fleeter than Eclipse? Can our prize-cattle and sheep be still further improved? Will a gooseberry ever weigh more than that produced by "London" in 1852? Will the beet-root in France yield a greater percentage of sugar? Will future varieties of wheat and other grain produce heavier crops than our present varieties? These questions cannot be positively answered; but it is certain that we ought to be cautious in answering by a negative. In some lines of variation the limit has probably been reached. Youatt believes that the reduction of bone in some of our sheep has already been carried so far that it entails great delicacy of constitution.[61] But seeing the great improvement within recent times in our cattle and sheep, and especially in our pigs; seeing the wonderful increase in weight in our poultry of all kinds during the last few years; he would be a bold man who would assert that perfection has been reached. Eclipse perhaps may never be beaten until all our race-horses have been rendered swifter, through the selection of the best horses during many generations; and then the old Eclipse may possibly be eclipsed; but, as Mr. Wallace has remarked, there must be an ultimate limit to the fleetness of every animal, whether under nature or domestication; and with the horse this limit has perhaps been reached. Until our fields are better manured, it may be impossible for a new variety of wheat to yield a heavier crop. But in many cases those who are best qualified to judge do not believe that the extreme point has as yet been reached even with respect to characters which have already been carried to a high standard of perfection. For instance, the short-faced tumbler-pigeon has been greatly modified; nevertheless, according to Mr. Eaton,[62] "the field is still as open for fresh competitors as it was one hundred years ago." Over and over again it has been said that

[61] Youatt on Sheep, p. 521. [62] 'A Treatise on the Almond Tumbler,' p. i.

perfection had been attained with our flowers, but a higher standard has soon been reached. Hardly any fruit has been more improved than the strawberry, yet a great authority remarks,[63] "it must not be concealed that we are far from the extreme limits at which we may arrive."

Time is an important element in the formation of our domestic races, as it permitts innumerable individuals to be born, and these when exposed to diversified conditions are rendered variable. Methodical selection has been occasionally practised from an ancient period to the present day, even by semi-civilised people, and during former times will have produced some effect. Unconscious selection will have been still more effective; for during a lengthened period the more valuable individual animals will occasionally have been saved, and the less valuable neglected. In the course, also, of time, different varieties, especially in the less civilised countries, will have been more or less modified through natural selection. It is generally believed, though on this head we have little or no evidence, that new characters in time become fixed; and after having long remained fixed it seems possible that under new conditions they might again be rendered variable.

How great the lapse of time has been since man first domesticated animals and cultivated plants, we begin dimly to see. When the lake-buildings of Switzerland were inhabited during the Neolithic period, several animals were already domesticated and various plants cultivated. If we may judge from what we now see of the habits of savages, it is probable that the men of the earlier Stone period—when many great quadrupeds were living which are now extinct, and when the face of the country was widely different from what it now is—possessed at least some few domesticated animals, although their remains have not as yet been discovered. If the science of language can be trusted, the art of ploughing and sowing the land was followed, and the chief animals had been already domesticated, at an epoch so immensely remote, that the Sanskrit, Greek, Latin, Gothic, Celtic, and Sclavonic languages had not as yet diverged from their common parent-tongue.[64]

[63] M. J. de Jonghe, in 'Gard. Chron.,' 1858, p. 173.

[64] Max. Müller, 'Science of Language,' 1861, p. 223.

It is scarcely possible to overrate the effects of selection occasionally carried on in various ways and places during thousands of generations. All that we know, and, in a still stronger degree, all that we do not know,[65] of the history of the great majority of our breeds, even of our more modern breeds, agrees with the view that their production, through the action of unconscious and methodical selection, has been almost insensibly slow. When a man attends rather more closely than is usual to the breeding of his animals, he is almost sure to improve them to a slight extent. They are in consequence valued in his immediate neighbourhood, and are bred by others; and their characteristic features, whatever these may be, will then slowly but steadily be increased, sometimes by methodical and almost always by unconscious selection. At last a strain, deserving to be called a sub-variety, becomes a little more widely known, receives a local name, and spreads. The spreading will have been extremely slow during ancient and less civilised times, but now is rapid. By the time that the new breed had assumed a somewhat distinct character, its history, hardly noticed at the time, will have been completely forgotten; for, as Low remarks,[66] "we know how quickly the memory of such events is effaced."

As soon as a new breed is thus formed, it is liable through the same process to break up into new strains and sub-varieties. For different varieties are suited for, and are valued under, different circumstances. Fashion changes, but, should a fashion last for even a moderate length of time, so strong is the principle of inheritance, that some effect will probably be impressed on the breed. Thus varieties go on increasing in number, and history shows us how wonderfully they have increased since the earliest records.[67] As each new variety is produced, the earlier, intermediate, and less valuable forms will be neglected, and perish. When a breed, from not being valued, is kept in small numbers, its extinction almost inevitably follows sooner or later, either from accidental causes of destruction or from close interbreeding; and this is an event which, in the case of well-marked breeds, excites attention. The birth or production of a new domestic race is so slow a process that it

[65] Youatt on Cattle, pp. 116, 128. [66] 'Domesticated Animals,' p. 188.
[67] Volz, 'Beiträge zur Kulturgeschichte,' 1852, s. 99 *et passim*.

escapes notice; its death or destruction is comparatively sudden, is often recorded, and when too late sometimes regretted.

Several authors have drawn a wide distinction between artificial and natural races. The latter are more uniform in character, possessing in a high degree the character of natural species, and are of ancient origin. They are generally found in less civilised countries, and have probably been largely modified by natural selection, and only to a small extent by man's unconscious and methodical selection. They have, also, during a long period, been directly acted on by the physical conditions of the countries which they inhabit. The so-called artificial races, on the other hand, are not so uniform in character; some have a semi-monstrous character, such as "the wry-legged terriers so useful in rabbit-shooting,"[68] turnspit dogs, ancon sheep, niata oxen, Polish fowls, fantail-pigeons, &c.; their characteristic features have generally been acquired suddenly, though subsequently increased in many cases by careful selection. Other races, which certainly must be called artificial, for they have been largely modified by methodical selection and by crossing, as the English race-horse, terrier-dogs, the English game-cock, Antwerp carrier-pigeons, &c., nevertheless cannot be said to have an unnatural appearance; and no distinct line, as it seems to me, can be drawn between natural and artificial races.

It is not surprising that domestic races should generally present a different aspect from natural species. Man selects and propagates modifications solely for his own use or fancy, and not for the creature's own good. His attention is struck by strongly marked modifications, which have appeared suddenly, due to some great disturbing cause in the organisation. He attends almost exclusively to external characters; and when he succeeds in modifying internal organs,—when for instance he reduces the bones and offal, or loads the viscera with fat, or gives early maturity, &c.,—the chances are strong that he will at the same time weaken the constitution. On the other hand, when an animal has to struggle throughout its life with many competitors and enemies, under circumstances inconceivably complex and liable to change, modifications of the most varied nature—in the internal organs as well as in external characters, in the

[68] Blaine, 'Encyclop. of Rural Sports,' p. 213.

functions and mutual relations of parts—will be rigorously tested, preserved, or rejected. Natural selection often checks man's comparatively feeble and capricious attempts at improvement; and if this were not so, the result of his work, and of nature's work, would be even still more different. Nevertheless, we must not overrate the amount of difference between natural species and domestic races; the most experienced naturalists have often disputed whether the latter are descended from one or from several aboriginal stocks, and this clearly shows that there is no palpable difference between species and races.

Domestic races propagate their kind far more truly, and endure for much longer periods, than most naturalists are willing to admit. Breeders feel no doubt on this head; ask a man who has long reared Shorthorn or Hereford cattle, Leicester or Southdown sheep, Spanish or Game poultry, tumbler or carrier-pigeons, whether these races may not have been derived from common progenitors, and he will probably laugh you to scorn. The breeder admits that he may hope to produce sheep with finer or longer wool and with better carcases, or handsomer fowls, or carrier-pigeons with beaks just perceptibly longer to the practised eye, and thus be successful at an exhibition. Thus far he will go, but no farther. He does not reflect on what follows from adding up during a long course of time many, slight, successive modifications; nor does he reflect on the former existence of numerous varieties, connecting the links in each divergent line of descent. He concludes, as was shown in the earlier chapters, that all the chief breeds to which he has long attended are aboriginal productions. The systematic naturalist, on the other hand, who generally knows nothing of the art of breeding, who does not pretend to know how and when the several domestic races were formed, who cannot have seen the intermediate gradations, for they do not now exist, nevertheless feels no doubt that these races are sprung from a single source. But ask him whether the closely allied natural species which he has studied may not have descended from a common progenitor, and he in his turn will perhaps reject the notion with scorn. Thus the naturalist and breeder may mutually learn a useful lesson from each other.

Summary on Selection by Man.—There can be no doubt that

methodical selection has effected and will effect wonderful results. It was occasionally practised in ancient times, and is still practised by semi-civilised people. Characters of the highest importance, and others of trifling value, have been attended to, and modified. I need not here repeat what has been so often said on the part which unconscious selection has played: we see its power in the difference between flocks which have been separately bred, and in the slow changes, as circumstances have slowly changed, which many animals have undergone in the same country, or when transported into a foreign land. We see the combined effects of methodical and unconscious selection in the great amount of difference between varieties in those parts or qualities which are valued by man, in comparison with those which are not valued, and consequently have not been attended to. Natural selection often determines man's power of selection. We sometimes err in imagining that characters, which are considered as unimportant by the systematic naturalist, could not be affected by the struggle for existence, and therefore be acted on by natural selection; but striking cases have been given, showing how great an error this is.

The possibility of selection coming into action rests on variability; and this is mainly caused, as we shall hereafter see, by changes in the conditions of life. Selection is sometimes rendered difficult, or even impossible, by the conditions being opposed to the desired character or quality. It is sometimes checked by the lessened fertility and weakened constitution which follow from long-continued close interbreeding. That methodical selection may be successful, the closest attention and discernment, combined with unwearied patience, are absolutely necessary; and these same qualities, though not indispensable, are highly serviceable in the case of unconscious selection. It is almost necessary that a large number of individuals should be reared; for thus there will be a fair chance of variations of the desired nature arising, and every individual with the slightest blemish or in any degree inferior may be freely rejected. Hence length of time is an important element of success. Thus, also, propagation at an early age and at short intervals favours the work. Facility in pairing animals, or their inhabiting a confined area, is advantageous as a check to free crossing. Whenever and wherever selection is not practised, distinct races are not formed. When any one part of the body or quality is not attended to, it remains either unchanged or varies in a fluctuating manner, whilst at the same time other parts and other qualities may become permanently and greatly modified. But from the tendency to reversion and to continued variability, those parts or organs which are now undergoing rapid improvement through selection, are likewise found to vary much. Consequently highly-bred animals, when neglected, soon degenerate; but we have no reason to believe that the effects of long-continued selection would, if the conditions of life remained the same, be soon and completely lost.

Man always tends to go to an extreme point in the selection, whether methodical or unconscious, of all useful and pleasing qualities. This is an important principle, as it leads to continued divergence, and in some rare cases to convergence of character. The possibility of continued divergence rests on the tendency in each part or organ to go on varying in the same manner in which it has already varied; and that this occurs, is proved by the steady and gradual improvement of many animals and plants during lengthened periods. The principle of divergence of character, combined with the neglect and final extinction of all previous, less-valued, and intermediate varieties, explains the amount of difference and the distinctness of our several races. Although we may have reached the utmost limit to which certain characters can be modified, yet we are far from having reached, as we have good reason to believe, the limit in the majority of cases. Finally, from the difference between selection as carried on by man and by nature, we can understand how it is that domestic races often, though by no means always, differ in general aspect from closely allied natural species.

Throughout this chapter and elsewhere I have spoken of selection as the paramount power, yet its action absolutely depends on what we in our ignorance call spontaneous or accidental variability. Let an architect be compelled to build an edifice with uncut stones, fallen from a precipice. The shape of each fragment may be called accidental; yet the shape of each has been determined by the force of gravity, the nature of the rock, and the slope of the precipice,—events and circumstances, all of which depend on natural laws; but there is no relation between these laws and the purpose for which each fragment is used by the builder. In the same manner the variations of each creature are determined by fixed and immutable laws; but these bear no relation to the living structure which is slowly built up through the power of selection, whether this be natural or artificial selection.

If our architect succeeded in rearing a noble edifice, using the rough wedge-shaped fragments for the arches, the longer stones for the lintels, and so forth, we should admire his skill even in a higher degree than if he had used stones shaped for the purpose. So it is with selection, whether applied by man or by nature; for though variability is indispensably necessary, yet, when we look at some highly complex and excellently adapted organism, variability sinks to a quite subordinate position in importance in comparison with selection, in the same manner as the shape of each fragment used by our supposed architect is unimportant in comparison with his skill.

Part IV

HUMAN EVOLUTION BY ARTIFICIAL SELECTION

Editor's Comments on Papers 16 and 17

16 DARWIN
Excerpt from *General Summary and Concluding Remarks*

17 WYATT
John Humphrey Noyes and the Stirpicultural Experiment

Charles Darwin, in *The Descent of Man,* used the analogy between the selective breeding of domesticated animals and human reproduction to decry the adverse effects of many social practices on the action of natural selection in civilized nations (Darwin, 1871:167–180) and to contend that man "might by selection do something not only for the bodily constitution and frame of his offspring, but for their intellectual and moral qualities" (Darwin, 1871; see also Paper 16). Francis Galton, Charles Darwin's cousin, had contended:

> If a twentieth part of the cost and pains were spent in measures for the improvement of the human race that is spent on the improvement of the breed of horses and cattle, what a galaxy of genius might we not create! . . . Men and women of the present day are, to those we might hope to bring into existence, what the pariah dogs of an Eastern town are to our own highly-bred varieties. (Galton, 1865:165–166)

Francis Galton's proposals to bring about the evolution of a gifted race of human beings by artificial selection was considered to be very utopian by Charles Darwin (Darwin and Seward, 1903). While Darwin did not make any specific proposals for the social control of human reproduction for eugenic purposes, he did contend that in Victorian industrial society, "There should be open competition for all men; and the most able should not be prevented by law or customs from succeeding best and rearing the largest number of offspring" (Darwin, 1871, II:403). Darwin also recorded his negative views concerning the possible eugenic role that birth control could play in directing human evolution in his private correspondence with George Gaskell (Clapperton, 1885). The Benchmark volume *Eugenics: Then and Now,* edited by Carl Bajema (1976), contains reprints of papers by Francis Galton, John Humphrey Noyes, H. J. Muller, Julian Huxley, and others who advocated the development and adoption of eugenics, the social regulation of human reproduction to bring about the genetic improvement of the human race.

The essay "Scientific Propagation" by John Humphrey Noyes (1870) contains, first of all, quotations from numerous writers on the analogy between the selective breeding of animals and human reproduction; second, a review of "what has been done for plants and animals" and "how it was done"; and third, Noyes's views on "how far and by what means can the same be done for human beings." Noyes was responsible for the experiment in the artificial selective breeding of human beings that was carried out by the Christian communist Oneida Community between 1868 and 1879 (Noyes and Noyes, 1923). Philip Wyatt's "John Humphrey Noyes and the Stirpicultural Experiment" (Paper 17) is the most recent review of the Oneida Community experiment in eugenics. A critical historical analysis of the development of the Oneida Community's complex marriage system and its stirpicultural experiment can be found in *Sex and Marriage in Utopian Communities: 19th Century America* (Muncy, 1973:160–196).

The rate at which artificial selection can bring about the spread of genes in populations of domesticated cattle, horses, and sheep has been greatly increased by the use of artificial insemination in selective breeding. Artificial insemination was first used extensively in selective breeding by the Russians at the turn of this century (Rudduck, 1948), and by the Americans, English, and West Europeans after World War I (Robertson, 1954).

The use of artificial insemination to achieve eugenic change in the human species was championed by the American geneticist, H. J. Muller (1935, 1959; see Carlson, 1973; Bajema, 1976). An estimated 10,000 births occur each year in the United States as the consequence of artificial insemination by donor sperm (Curie-Cohen, Luttrell, and Shapiro, 1979). Past artificial selection of human sperm donors on the basis of eugenic criteria has already had eugenic consequences. A follow-up study of children conceived by artificial insemination found that such children developed higher IQs (Iizuki et al., 1968). Physicians' use of more stringent eugenic criteria in the artificial selection of donor human gametes as they employ AID (artificial insemination using donor sperm), ETD (egg or embryo transfer using a donor egg), or ETDD (both AID and ETD) technology would probably bring about an even greater increase in the genes involved in the development of good health and mental qualities that can be attributed to the use of AID and ETD technology (Smith, 1976).

16

Reprinted from pages 402–405 of *The Descent of Man, and Selection in Relation to Sex*, vol. II, John Murray, London, 1871

GENERAL SUMMARY AND CONCLUDING REMARKS

C. R. Darwin

[*Editor's Note:* In the original, material precedes this excerpt.]

Man scans with scrupulous care the character and pedigree of his horses, cattle, and dogs before he matches them ; but when he comes to his own marriage he rarely, or never, takes any such care. He is impelled by nearly the same motives as are the lower animals when left to their own free choice, though he is in so far superior to them that he highly values mental charms and virtues. On the other hand he is strongly attracted by mere wealth or rank. Yet he might by selection do something not only for the bodily constitution and frame of his offspring, but for their intellectual and moral qualities. Both sexes ought to refrain from marriage if in any marked degree inferior in body or mind; but such hopes are Utopian and will never be even partially realised until the laws of inheritance are thoroughly known. All do good service who aid towards this end. When the principles of breeding and of inheritance are better understood, we shall not hear ignorant members of our legislature rejecting with scorn a plan for ascertaining by an easy method whether or not consanguineous marriages are injurious to man.

The advancement of the welfare of mankind is a most intricate problem: all ought to refrain from marriage who cannot avoid abject poverty for their children; for poverty is not only a great evil, but tends to its own increase by leading to recklessness in marriage. On the other hand, as Mr. Galton has remarked, if the prudent avoid marriage, whilst the reckless marry, the inferior members will tend to supplant the better members of society. Man, like every other animal, has no doubt advanced to his present high condition through a struggle for existence consequent on his rapid multiplication; and if he is to advance still higher he must remain subject to a severe struggle. Otherwise he would soon sink into indolence, and the more highly-gifted men would not be more successful in the battle of life than the less gifted. Hence our natural rate of increase, though leading to many and obvious evils, must not be greatly diminished by any means. There should be open competition for all men; and the most able should not be prevented by laws or customs from succeeding best and rearing the largest number of offspring. Important as the struggle for existence has been and even still is, yet as far as the highest part of man's nature is concerned there are other agencies more important. For the moral qualities are advanced, either directly or indirectly, much more through the effects of habit, the reasoning powers, instruction, religion, &c., than through natural selection; though to this latter agency the social instincts, which afforded the basis for the development of the moral sense, may be safely attributed.

The main conclusion arrived at in this work, namely that man is descended from some lowly-organised form, will, I regret to think, be highly distasteful to many persons. But there can hardly be a doubt that we are descended from barbarians. The astonishment which I felt on first seeing a party of Fuegians on a wild and broken shore will never be forgotten by me, for the reflection at once rushed into my mind—such were our ancestors. These men were absolutely naked and bedaubed with paint, their long hair was tangled, their mouths frothed with excitement, and their expression was wild, startled, and distrustful. They possessed hardly any arts, and like wild animals lived on what they could catch; they had no government, and were merciless to every one not of their own small tribe. He who has seen a savage in his native land will not feel much shame, if forced to acknowledge that the blood of some more humble creature flows in his veins. For my own part I would as soon be descended from that heroic little monkey, who braved his dreaded enemy in order to save the life of his keeper; or from that old baboon, who, descending from the mountains, carried away in triumph his young comrade from a crowd of astonished dogs—as from a savage who delights to torture his enemies, offers up bloody sacrifices, practises infanticide without remorse, treats his wives like slaves, knows no decency, and is haunted by the grossest superstitions.

Man may be excused for feeling some pride at having risen, though not through his own exertions, to the very summit of the organic scale; and the fact of his having thus risen, instead of having been aboriginally placed there, may give him hopes for a still higher destiny in the distant future. But we are not here concerned with hopes or fears, only with the truth as far as our reason allows us to discover it. I have given the evidence to the best of my ability; and we must acknowledge, as it seems to me, that man with all his noble qualities, with sympathy which feels for the most debased, with benevolence which extends not only to other men but to the humblest living creature, with his god-like intellect which has penetrated into the movements and constitution of the solar system—with all these exalted powers—Man still bears in his bodily frame the indelible stamp of his lowly origin.

17

Reprinted from *J. Hist. Med. Allied Sci.* **31**:55–66 (1976)

John Humphrey Noyes and the Stirpicultural Experiment

PHILIP R. WYATT

WE ARE opposed to excessive, and of course, oppressive procreation, which is almost universal. We are opposed to random procreation, which is unavoidable in the marriage system. But we are in favor of intelligent well-ordered procreation. The physiologists say that the race cannot be raised from ruin until propagation is made a matter of science; but they point out no way of making it so. Procreation is controlled and reduced to a science in the case of valuable domestic brutes; but marriage and fashion forbid any such system among human beings. We believe the time will come when involuntary and random propagation will cease, and when scientific combination will be applied to human generation as freely and successfully as it is to that of other animals. They will be open for this when amativeness can have its proper gratification without drawing after it procreation as a necessary sequence. And at all events, we believe that good sense and benevolence will very soon sanction and enforce the rule that women shall bear children only when they choose. They have the principal burdens of breeding to bear, and they rather than men should have their choice of time and circumstances at least till science takes charge of the business.[1]

This paragraph, which strongly expressed the speaker's belief in the application of eugenic management to the human population, was written in 1849. The author of it was the leader of a group of individuals who succeeded in implementing controlled matings where scientific scrutiny 'took charge of the business' of determining which individuals should be permitted to produce children.

Discussion in favor of eugenic management and suggestions of pro-

1. R. A. Parker, *A Yankee saint: John Humphrey Noyes and the Oneida Community* (New York, 1935), p. 253.

The author thanks Dr. W. L. Parker for providing the opportunity for this work to be carried out; Miss H. B. Lillie for her aid in the preparation of this manuscript; and the staff of Deer Lodge Hospital for providing an atmosphere which permitted this work to be completed.

posed eugenic policies are today countered with valid scientific criticism, emotional harangues, and the spectre of the holocaust which occurred under the German Third Reich. One example of a society which succeeded in the application of eugenic policies without creating chaos in its social structure has escaped the attention of most eugenicists and their opponents. This society composed of the followers of John Humphrey Noyes, was called the Oneida Community.[2]

Few societies have initiated policies of planned matings of its members with the production of healthier offspring as their goal. The Oneida Community was a group of individuals who implemented a policy of stirpiculture, the production of individuals by controlled mating of selected members of the community. Their policy of stirpiculture represented one of the few experiments in human eugenics where records still exist.[3] Stirpiculture was defined as 'selective breeding for the development of strains with certain characteristics.' In the Oneida Community, where the term was coined, it referred to selective breeding for the development of a human race with moral and physical perfection.[4]

The Oneida Community was founded by John Humphrey Noyes at Putney, Vermont, in 1841 with approximately fifty members. By 1848, when the community moved to Oneida, New York, there were approximately ninety followers of Noyes' teachings which were based on the principles of Perfectionism.[5]

The early nineteenth century was a period of great upheaval against the established churches. Various splinter religious groups were formed, one of which was a group known as Perfectionists or Utopians, who believed that man could become perfect on earth. Their beliefs ranged from a total isolation of the community from the rest of society to total abstinence from moral and physical sins. As a result of the range of beliefs, numerous small groups developed who called themselves Perfectionists. There was continuous effort to achieve leadership of these small groups. One man who succeeded in achieving leadership of such a group was John Humphrey Noyes.

Noyes was born 3 September 1811 and was educated at Dartmouth, Andover, and Yale, where he was graduated in 1833 with a degree in theology. He was caught up in the Perfectionism movement and in 1833

2. A. N. McGee, 'An experiment in human stirpiculture,' *Amer. Anthro.*, 1891, *4*, 319–325.
3. M. L. Carden, *Oneida: utopian community to modern corporation* (Baltimore, 1969), p. 64.
4. *Webster's new twentieth century dictionary*, 2nd ed., s.v. 'stirpiculture.'
5. G. W. Noyes, *John Humphrey Noyes, the Putney Community* (Oneida, 1931), pp. 1–15.

joined a free church at New Haven to develop further his beliefs and doctrines. In 1834, Noyes was asked to leave the New Haven free church when he declared himself sinless.[6] From that point on he sought to form his own congregation of Perfectionists. Noyes' interpretation of Perfectionism was based on his belief that earthly perfection had been possible for man to achieve since 70 A.D. which represented the completion of the first coming of Christ and the death of His apostles. The second coming of Christ was possible any time after Christ's death, but the majority of His apostles had strayed from the path of true perfection and their imperfection was reflected in the doctrines of the nineteenth century churches which were based on the interpretation of Christ's teaching by these same apostles. Noyes believed that the second coming of Christ occurred when one's soul realized Christ's presence. The realization of Christ's presence gave a spiritual sense of perfection which would help the soul to gain power over the physical body and perfect it.[7] Noyes gradually gained a position of authority in the Perfectionist movement, and he began to attract followers who located at Putney, Vermont. They signed a contract of partnership and formed the Putney Corporation.

Noyes and his followers believed that the self-perfection they each sought progressed through four stages. First, one achieved a spiritual perfection, which was followed by an intellectual perfection, then one strove toward moral and, finally, physical perfection. Noyes hoped to develop a Perfectionist group which would achieve these four stages of development successfully, and he made careful efforts to lead the community members into the successive stages of perfection. Noyes' desires for perfection in the members produced at least four unique concepts which community members were required to adopt in order to achieve their goal. These were male continence, complex marriage, community child care, and the stirpicultural experiment. The practice of male continence and complex marriage, which were initiated at Putney, led to the expulsion of the community from Putney and its subsequent migration to Oneida, New York.

The first of the four principles was male continence, or *coitus reservatus*.[8] The practice of male continence was essential as a method of birth control in a community which permitted sexual communism. Male continence freed women from family responsibilities and allowed them to function as

6. Carden, *Oneida* (n. 3), pp. 2–4.
7. *Ibid.*, p. 13.
8. R. S. Fogarty, *American utopianism* (Ithaca, 1972), p. 80.

full members of the Oneida Community in all social, economic, and physical activities. Noyes aptly described male continence as

> The situation (male continence) may be compared to a stream in three conditions, viz., 1. a fall; 2. a course of rapids above the fall; and 3. still water above the rapids. The skilful boatman may choose whether he will remain in the still water, or venture more or less down the rapids, or run his boat over the fall. But there is a point on the verge of the fall where he has no control over his course; and just above that there is a point where he will have to struggle with the current in a way which will give his nerves a severe trial, even though he may escape the fall. If he is willing to learn, experience will teach him the wisdom of continuing his excursions to the region of easy rowing, unless he has an object in view that is worth the cost of going over the fall.[9]

Noyes believed that the practice of male continence not only functioned as a means of birth control, but also served an alternate function of providing a healthier, stronger breed of children.

> Male continence not only relieves us of undesirable propagation but opens the way for scientific propagation. We are not opposed to procreation... but we are in favor of intelligent, well-ordered procreation. The time will come when scientific combination will be freely and successfully applied to human generation. The common objection to Male Continence is that it is unnatural and unauthorized by the example of the lower animals. But cooking, wearing clothes, living in houses are unnatural in the same sense. In a higher sense I believe it is natural for rational beings by invention and discovery to improve nature. Until men and women by moral cultures elevate their sexual life above that of the brutes they are living in unnatural degradation.[10]

Male continence was advocated by Noyes in 1846 and continued to be practiced by all members of the community until 1880 when the community disbanded.

The success of male continence was obvious in the community. Over a period of twenty-three years from 1846 to 1869, there were only forty-four births: eight births to women who were pregnant prior to joining the community, five births which were sanctioned by the group, and the remaining thirty-one births which represented accidental conceptions in a population of approximately 200 adults.

The success of male continence in freeing the women from unwanted pregnancies permitted the practice of complex marriage in the commu-

9. E. Van de Warken, 'A gynecological study of the Oneida Community,' *Amer. J. Obstet. Gyn.*, 1884, *17*, 785–810.

10. Noyes, *Noyes* (n. 5), pp. 114–115.

nity's effort to reach a more perfect status. In complex marriage, all adult members of the community participated as a large married family sharing all the economic, social, cultural, and sexual benefits of such an arrangement. This practice led to the community's expulsion from Putney because it was felt by the neighboring areas that the Perfectionists were advocates of free love and were opposed to marriage, and thus were dangerous to the morals of the surrounding communities. Noyes made stringent efforts to clarify the difference between free love and complex marriage with a detailed description.

> The essential differences between marriage and free love may be stated thus: Marriage is a permanent union; Free Love is a hireling system. Marriage makes a man responsible for his acts to a woman; Free Love allows him to impose his will and go his way without responsibility. Marriage provides for the maintenance and education of children; Free Love ignores and leaves them to chance. In respect to every one of these points we stand with marriage. Free Love with us does not mean freedom to love today and leave tomorrow; nor freedom to take a woman's person and keep our property to ourselves; nor freedom to freight a woman with our offspring and send her downstream without care or help, nor freedom to beget children and leave them to the street and the poorhouse.
>
> Our communities are as distinctly bounded and separated from outside society as ordinary families; the tie that binds us together is as permanent and sacred to say the least as that of marriage, for it is our religion; we receive no new members (except by mistake) who do not give heart and hand to the Community for life and forever. Whoever will take the trouble to follow our track from the beginning will find no forsaken women or children by the way.[11]

Complex marriage was approved for the community members in 1846. Prior to this time Noyes and a circle of founding members were the only individuals who were permitted to utilize sexual communism.

Complex marriage was started under strict regulation. The women were free to reject all suitors, and the sexual encounter was normally made through a third party whose function it was to record all such encounters for preservation in the community records, which it was felt would limit the chance of an exclusive attachment occurring between a man and woman.

Noyes was concerned lest complex marriage would be seen by outsiders as the destruction of social order. He clearly stated the difference

11. C. N. Robertson, *Oneida Community: an autobiography, 1851–1876* (Syracuse, 1970), p. 282.

between complex marriage and free love but was unsure that this explanation was enough. He therefore outlined his future plans for the community and demonstrated that complex marriage was essential to the improvement of mankind.

> The institutions that shall at some future time supersede marriage and its accessories, whatever may be their details, must include certain essentials, negative and positive, which can be foreseen now with entire certainty.
>
> In the first place, they must not lessen human liberty. Here we touch the main point of the difference between the cases of animals and men, and the point of difficulty for our whole problem. . . . Man as a race has no visible superior. The fact declares that his destiny is self-government. . . . The liberty already won must not be diminished, but increased. If there is to be suppression, it must not be by castration and confinement, as in the case of animals, or even by law and public opinion, as men are now controlled, but by the free choice of those who love science well enough 'to make themselves eunuchs for the Kingdom of Heaven's sake.'[12]

The third principle, which was a major step toward the attainment of total perfection by the community and toward the stirpicultural experiment, was the policy of community child care. The survival of the community depended on Noyes' ability to create a strong economic base. This he achieved by developing a successful farming industry, a silk thread factory, and a trap manufacturing plant which eventually became the highly successful Oneida silverware company. The success of the child care program was dependent upon an economy capable of supporting it. Children were present at Oneida through the addition of new members to the community and the birth of a small number of infants. In 1849 the community built the Children's Home. From that time the children in the community followed a prescribed routine. Infants were cared for by their mothers until about nine months; up to approximately eighteen months the mother had charge of the infant only at night, while during the day the child was cared for in the Children's Home. After the child was weaned, the mother no longer had an individual responsibility for the child.[13]

In the Children's Home, the children were under the care of three men and fifteen women until they reached adolescence, at which time they were accepted as adult members of the community. The children were removed from the influence of their parents and placed under the care of community members. The environment in the Children's Home offered

12. Parker, *Yankee saint* (n. 1), p. 256.
13. Robertson, *Oneida Community* (n. 11), pp. 311–334.

its occupants instruction in school work, supervision during the day, and involvement for the older children in the normal day-to-day routine of the community. This environment represented the community's concept of the best possible opportunities for development.

The child care program was the community's attempt to indoctrinate the children in the Perfectionist philosophy. The boys aged twelve or older were involved in running the various commercial enterprises of the community, and the girls aged ten or older were assigned household duties. Since all toys belonged to a large collection from which the children drew, the concept of communal ownership was introduced at an early age. The children were indoctrinated in the Perfectionist philosophy in a healthy country environment, in which love and attention were provided by all the members of the community. Noyes believed this system was superior to the traditional rearing of children in the family unit. Noyes and the other members of the Oneida Community recognized that culture was a tremendously potent instrument for adaptation to the environment. The community strove to provide all its children with the greatest possible cultural advantages which they felt would be provided by a central care agency, and could not be provided by individual family units.

With the establishment of a male continence, complex marriage, and community child care, the final step in Noyes' plan for a perfect race could begin.

> The subject of bearing and rearing children, though second to none in interest, has been for various considerations postponed until we should arrive at it in the due order of things. Childbearing, when it is undertaken, should be a voluntary affair, one in which the choice of the mother, and the sympathy of all good influences should concur. Our principles accord to woman a just and righteous freedom in this particular, and however strange such an idea may seem now, the time cannot be distant when any other idea or practice will be scorned as essential barbarism.[14]

To assure that the children borne by the women of the community were as perfect as possible, Noyes formulated the policy of stirpiculture, or the culture of a new race, a race produced by carefully controlled selective breeding. Noyes was attempting to produce an improvement in the human race by prohibiting the breeding of 'inferior' individuals (who were

14. *Ibid.*, p. 341.

not able to achieve some form of self-perfection) and by endorsing the propagation of children from the best individuals the community offered. He then hoped to interbreed these offspring, with the occasional introduction of outsiders, to produce the best possible individuals. This was an effort to achieve scientific breeding in humans, utilizing the principles of nineteenth century animal breeding.

> Assuming then that we have to deal with a science in breeding that gives definite results, we refer again to our question, what is the point to be first aimed at in the improvement of the human stock? Shall it be physical perfection, beauty of form, strength, complexion, health? Shall it be sagacity, acuteness of mind? Shall it be amiability of mind? These are questions for consideration. Again, shall we adopt some fixed type, as the Anglo-Saxon man, or the classical Greek type, and selecting the most beautiful examples of some of these classes, breed to them as a standard? Or shall we recognize the variety of nature as the rule, and seek only to perfect the multifarious types that are now extant, each according to its own peculiarities of style and constitution? These points are to be resolved by careful thought. The subject is new, and will have to be approached by degrees, until practical experiment shall have thrown its light upon the broad pathway that, through truthful, scientific propagation, must lead the race up to its ideal development and destiny.[15]

The community unanimously voted to adopt the policy of stirpiculture. To provide Noyes a base stock with which he could work, fifty-three women signed pledges dedicating themselves to the project.

> 1. That we do not belong to ourselves in any respect, but that we do belong first to God, and second to Mr. Noyes as God's true representative.
>
> 2. That we have no rights or personal feelings in regard to child bearing which shall in the least degree oppose or embarrass him in his choice of scientific combinations.
>
> 3. That we will put aside all envy, childishness and self-seeking, and rejoice with those who are chosen candidates; that we will, if necessary, become martyrs to science, and cheerfully resign all desire to become mothers, if for any reason Mr. Noyes deem us unfit material for propagation. Above all, we offer ourselves 'living sacrifices' to God and true Communism.[16]

Thirty-eight of the men of the community signed a similar document declaring their willingness to participate in the experiment.[17] Thus by 1868 Noyes was in a position to implement his policy of stirpiculture.

15. *Ibid.*, p. 342.
16. Carden, *Oneida* (n. 3), p. 62.
17. *Ibid.*, pp. 61–62.

From 1869 to 1879, fifty-eight live births occurred in the Oneida Community as a result of the stirpicultural experiments. The matings of the parents were all reviewed by Noyes and the central members of the community or the stirpicultural committee which existed for a period of fifteen months beginning in 1875, before any individuals were allowed to have children. A total of fifty-one applications for children were submitted, of which nine were rejected as unfit, and forty-two applications were accepted. The remaining matings were either solicited by the stirpiculture committee or resulted from accidental conceptions. Of the fifty-eight live births, there were no maternal losses, and only one child died, and that of an instrument delivery.[18]

COMPARISON OF DEATHS BETWEEN PRESTIRPICULTURAL AND STIRPICULTURAL POPULATIONS IN ONEIDA COMMUNITY

	No. of deaths observed	*No. of deaths expected (as cited in footnote 19)*	*No. of deaths expected (as cited in footnote 22)*	*No. of deaths expected (as cited in footnote 22)*
44 Prestirpicultural births	10	11	7	13
58 Stirpicultural births	1	18	9	18

The documents relating to the stirpiculture experiment were, for the most part, destroyed so that the specific factors Noyes used in the selection process are not available.[19] It is known that initially the selection process for the most suitable parents of the stirpicults was Lamarckian, based on the spiritual development of the parents, but by 1875 when the formal stirpicultural committee was formed, the physical and mental development of the parents played an important role in the selection process. The primary criterion in the philosophy of the stirpicultural experiment was that 'persons themselves must improve before they can transmit improvement to their offspring.'[20]

The matings which produced progeny were distributed evenly over the

18. Hilda H. Noyes and George W. Noyes, 'The Oneida experiment in stirpiculture' in International eugenics congress. 2d, New York, 1921. *Eugenics, genetics and the family* (Baltimore, 1923), pp. 374–386.

19. Robertson, *Oneida Community* (n. 11), p. 337.

20. Carden, *Oneida* (n. 3), p. 62.

ten-year span. Thirty-four infants were born between 1869 and 1875, twenty-four between 1875 and 1878, and none were born in 1879 as the practice of stirpiculture was discontinued during the two years prior to the community's disbandment. The parental generation of the stirpicults had unique characteristics. The fathers averaged 12.2 years older than the mothers. Thirty of the fathers sired only one child, while the remaining twenty-eight children were the offspring of ten fathers. The closest matings undertaken were two uncle-to-niece pairings.[21]

Comparison of the data available in the prestirpicults, the stirpicults, and the expected mortality of infants in the late nineteenth century is difficult due to the lack of data; but utilizing data published by H. H. Noyes in 1921 and Winslow and Holland in 1930, it can be demonstrated that the infant mortality rate in the 1800s was between 150 and 300 deaths per 1,000 live births.[22] Noyes suggested that the infant mortality rate in 1880 was approximately 225.[23] The table shows that ten of forty-four prestirpicults died under the age of five, which is apparently the expected number of deaths. Only one stirpicult died at birth, whereas between eight and eighteen deaths might have been expected statistically.

The information on infant mortality in mid-nineteenth century America is sparse and subject to errors, but there is a strong indication that a remarkably low mortality occurred among the children born in the stirpicultural experiment. The success of this experiment is suggested by the dramatic improvement in the health of the children produced. It is impossible, however, to credit this success to either the selective breeding process or the community care program. Many factors influence health and the Oneida Community changed a large number of factors in the production of the stirpicults so that conclusions supporting the importance of either selective breeding or community child care cannot be drawn from the stirpicult generation. The combination of selective breeding and community child care in the Oneida Community can be termed happy in its results, so far as the children were concerned.

The stirpiculture experiment terminated in 1879 when the community disbanded as a result of internal dissension and external pressure.[24] Only one generation of stirpicults was produced during the existence of the community. However, the principle was established in the minds of the

21. Noyes and Noyes, 'Oneida experiment' (n. 18), p. 378.
22. P. A. Harper, *Preventive pediatrics: child health and development* (New York, 1966), p. 591.
23. Noyes and Noyes, 'Oneida experiment' (n. 18), p. 386.
24. C. N. Robertson, *Oneida Community: the break-up 1876–1881* (Syracuse, 1972), p. 192.

experimenters that prenatal concern, as manifested in the controlled mating of the parents and postnatal care, as practiced in the Children's Home, had produced a healthy generation of children.

At the time of the breakup of the community, shares in its assets in excess of $550,000 were issued. Provision was made to provide educational facilities for all children under the age of sixteen until such time as they could claim their rightful shares. Several of these children went to college, several were successful in business, a large number were employed in the joint-stock company, Oneida Community Limited, one was killed by a train, and one fell from an airplane during the First World War. Two children were reported to be subnormal in development, one case being due to an accidental blow to the head at the age of seven, the other a suspected injury at birth; however, both were able to function independently. By the mid 1930s forty-six stirpicults were still alive and thirty-one of these were living at Oneida. The success of the stirpicults both in health and achievement was far above what might be expected on the basis of national averages, but once again this success cannot be attributed to the stirpiculture experiment or the community child care program separately, but to a product of the two.

In addition it is known that of the stirpicults fifty-five reached marriageable age. Of these fifty-five, five remained single, thirty-two married outside of the stirpiculture group, and eighteen of the stirpicults intermarried. Of the nine marriages where both parents were stirpicults, twenty-eight children were produced; none died under the age of five. In the thirty-two marriages where only one of the parents was a stirpicult, ninety-eight children were produced; four died under the age of five.[25]

Noyes may be considered a founder of the eugenics movement. He coined the term 'stirpiculture' twenty years before Galton proposed the term 'eugenics' in 1883.[26] By 1883 John Humphrey Noyes had controlled an eleven-year experiment in stirpiculture with fifty-eight individuals to show as demonstration of his theories that the nature of man could be improved by controlling both the environment in which man functioned, and his biological inheritance.

The Noyes experiment in stirpiculture thus succeeded in spirit, at least. A group of individuals, banded together in a communistic society, reached an understanding that their attributes—physical, spiritual, and emotional—as well as their failings must be controlled so that the children produced

25. Noyes and Noyes, 'Oneida experiment' (n. 18), p. 381.
26. F. Galton, *Inquiries to human faculty and its development* (London, 1883).

in their matings would be a reflection of their desire for perfection in the human race. They thought that the improvement of the community called for a belief in a distant abstract goal, that the improvement of the group was not a generational goal, but a goal for future generations. Their plans for the future led them to the belief that they must be concerned both with the cultural nature and the biological basis of man. Both had an equal footing in the Oneida Community's desire for more perfect offspring. It was truly a policy of eugenics.

Deer Lodge Hospital
Winnipeg, Manitoba

Part V

THE LIMITS OF SELECTION

Editor's Comments on Papers 18 Through 23

18 JOHANNSEN
Concerning Heredity in Populations and in Pure Lines

19 WELDON and PEARSON
Inheritance in Phaseolus vulgaris

20 YULE
Professor Johannsen's Experiments in Heredity: A Review

21 MacCURDY and CASTLE
Excerpts from *Selection and Cross-breeding in Relation to the Inheritance of Coat-pigments and Coat-patterns in Rats and Guinea Pigs*

22 PEARL and SURFACE
Is there a Cumulative Effect of Selection?

23 SURFACE
The Result of Selecting Fluctuating Variations. Data from the Illinois Corn Breeding Experiments

Charles Darwin was so impressed by the great "power of man in accumulating by selection successive slight variations" to modify breeds of domesticated animals that he contended:

> If selection consisted merely in separating some very distinct variety, and breeding from it, the principle would be so obvious as hardly to be worth notice; but its importance consists in the great effect produced by the accumulation in one direction, during successive generations, of differences absolutely inappreciable by an uneducated eye. . . . Not one man in a thousand has accuracy of eye and judgment sufficient to become an eminent breeder. (Darwin, 1859:32)

While Darwin admitted that the variations selected by the horticulturalists are often more abrupt (Darwin, 1859;32), he nevertheless contended: "As a general rule, I cannot doubt that the continued selection of slight variations, either in the leaves, the flowers, or the fruit, will produce races differing from each other chiefly in these characters" (Darwin, 1859:33). In *Origin* and in later publications, Charles Darwin contended that evolution occurs gradually as the result of selection operating on small individual differences rather than on "saltations," as large discontinuous variations then were often called (Vorzimmer, 1970). Charles Darwin adhered to the principle Natura non facit saltum ("Nature does not proceed by jumps") so strictly that his view was criticized by Alfred Wallace, Thomas Huxley, and Francis Galton, all of whom contended that evolution could occur by saltations as well as by small slight variations.

The biometricians led by Karl Pearson and W. F. R. Weldon championed Darwin's view that evolution involved continuous rather than discontinuous change. This set the stage for one of the bitterest intellectual battles ever to occur among evolutionists—the controversy between the Mendelians and the biometricians over the importance of selection as a factor of evolution during the first two decades of the twentieth century (Provine, 1971). The controversy centered on the question of whether evolution proceeded gradually or by jumps. Ernst Mayr has pointed out that both the Mendelian and biometrician schools of thought made the same three erroneous assumptions concerning the nature of variation:

1. There are two sharply distinguishable and quite different kinds of variation, discontinuous variation (sports, mutations) and continuous variation, a distinction going back to Darwin and earlier.
2. There is no distinction between changes in what we now call genotype and phenotype.
3. Individual and evolutionary variation are one and the same thing. (Mayr, 1973:146)

The research of William Bateson (1894) and Hugo de Vries (1901–1903) led them to champion the theory that large discontinuous variations were of paramount importance in the origin of species. de Vries observed approximately 800 discontinuous mutants belonging to seven different types among the 50,000 evening primrose (*Oenothera Lamarckiana*) plants he raised in his experimental garden (Dunn, 1965). Some of these mutants bred true and were considered to be new species by de Vries. This led de Vries (1901) to propose the mutation theory of evolution, the idea that new elementary species originated in a single step by saltation without selection playing a role. de Vries also argued that the limits of what could be achieved by selection of continuous variations would be reached within four or five generations.

The rediscovery of the Mendelian pattern of inheritance by Correns, de Vries, and Tschermak at the turn of the century was followed almost immediately by numerous reports that discontinuous variations but not continuous variations were inherited according to Mendelian principles. To make matters even worse, Karl Pearson and other biometricians who had developed statistical techniques to study the inheritance of continuous variation rejected Mendelian inheritance contending that it was either an unimportant or erroneous concept. The importance of selection generating gradual evolution by the accumulation of continuous variations was seriously challenged by such biologists as Wilhelm Johannsen, Herbert Jennings, and Raymond Pearl, all of whom had carried out artificial selection experiments (Provine, 1971).

The Danish botanist Wilhelm Johannsen (1857–1927) attempted to better understand the role of selection in evolution by studying the effects of selection on pure-lines. Each of Johannsen's pure-lines consisted of individuals who were all descendants of a single self-fertilized individual. Johannsen (1903, see also Paper 18) contended that the results of his artificial selection experiments on the weight of seeds in pure-lines of the princess bean, *Phaseolus vulgaris* (Paper 18), demonstrated that "fluctuating" (continuous) variability was not heritable and thus played an unimportant role in the evolution of pure-lines. Johannsen (1903:212) concluded: "Selection in populations acts in my cases only in so far as it chooses representatives of already existing types. These types are not successively formed, perhaps through the protection of those individuals which fluctuatingly vary in the specific direction; but they are found and isolated."

Johannsen argued that the parent-offspring correlation in his artificial selection experiment was negligible even though he had not calculated it. Johannsen's conclusions concerning the results of his selection experiments were criticized in 1903 by G. Yule (Paper 20) and by Karl Pearson and W. Weldon (Paper 19) who calculated the parent-offspring correlation to be a non-negligible +0.3481! Johannsen had admitted, "In several lines selection appears to have operated, in others the result was quite the opposite—on the whole, within the pure lines nothing at all has been achieved through selection" (Johannsen, 1903:206). Johannsen's research has been hailed as classic by numerous geneticists in spite of the fact that his published experimental data does not support his conclusions (see Roll-Hansen, 1978, 1980 for a defense of Johannsen's interpretation of the outcomes of his experiments.) Yet Johannsen's theory about selection and heredity in pure-lines (populations where each individual is supposed to have the same homozygous genotype) was basically correct (Provine, 1971).

Speaking at the Third International Congress of Genetics in 1906, Johannsen attempted to extend his pure-line theory of heredity to hybrid populations:

> we have no reason to suppose that an augmented fluctuation will be found in the new types which here may be formed by segregations and new combinations. Further research will, I have every conviction, give greater clearness as to the fundamental distinction of true *type differences* and *fluctuations*. (Johannsen, 1907:110)

Selection, according to Johannsen, was merely a process by which pure-lines (homozygous genotypes) were isolated from hybrid populations. Johannsen (1907:101) contended that the concept "that selection is able to shift a type in the same direction as that in which selection of its fluctuations is carried on" is absolutely erroneous. Johannsen argued that in the absence of new mutations, selection could not shift a hybrid population beyond the limits of variability exhibited in the F_2 generation. Johannsen's views on the limits to selection were very appealing to numerous students of Mendelian heredity who believed that mutations (as they then called discontinuous variations) played the only role in evolution (Provine, 1971). The failure of experiments involving artificial selection for size in Paramecium (Jennings, 1908, 1909, 1910) and for egg production in chickens (Pearl and Surface, 1909, Paper 22; Pearl, 1911*a*, 1911*b*, 1912) to bring about changes predicted by Darwinian selection theory were widely quoted as supporting Johannsen's pure-line theory and undermining Darwin's selection theory. Almost all Mendelians believed in the mutation theory of evolution at this time (Provine, 1971).

Yet by 1918 the effectiveness of Darwinian selection had been experimentally demonstrated, and the Mendelian theory of inheritance was shown to be complementary to Darwinian selection theory. Plant-breeding experiments conducted by H. Nilsson-Ehle (1909) on cereals and Edward East (1910) on maize provided evidence that genes behaving according to Mendelian theory could explain continuous variation. T. H. Morgan (1909:375) reported that he and his coworkers discovered "that some small variations are inherited" in *Drosophila* and that "if these be the material with which evolution is concerned, Darwin's assumptions in regard to the nature of variation will be, in part, justified." Mutations were shown to be involved in the inheritance of continuous variations as well as discontinuous variations. The mutationist theory of evolution championed by Hugo de Vries came under strong attack. The type of "mutations" in *Oenothera* that de Vries observed not only did not fit the typical pattern of inheritance for Mendelian genes but seemed to occur in only a few species at most. Fisher (1918) published a mathematical model showing how the inheritance of particulate Mendelian factors could account for the correlation between genetic relatives with respect to continuous variation.

The selective breeding experiments that William E. Castle (1867–1962) and his coworkers performed with guinea pigs and rats between 1901 and 1916 provided crucial evidence that selection operating on continuous variations could cause a character to change to a new level beyond the initial limits of variability exhibited by that character. Castle's first selective breeding experiment involved five generations of selection for an added digit in a population of guinea pigs into which an individual was born in 1901 that possessed a supernumerary fourth digit on its left hind foot. In his paper on "The Mutation Theory from the Standpoint of Animal Breeding," Castle (1905:522) contended that "this race was not produced by selection, though it was improved by that means." Castle placed more emphasis on the role of selection than mutation in his 1911 book *Heredity in Relation and Animal Breeding*:

> I have observed characters at first feebly manifested gradually improve under selection until they became established racial traits. Thus the extra toe of polydactylous guinea-pigs made its appearance as a poorly developed fourth toe on the left foot only. . . . Such individuals were selected throughout five successive generations, at the end of which time a good four-toed race had been established. (Castle 1911:120–121)

In 1907 Castle and MacCurdy published the results of five generations of two-way selection for the size of the dorsal stripe in hooded rats (excerpted as Paper 21). They succeeded in changing the shape of the hooded pattern in both directions by employing selective breeding to accumulate fluctuating variations. In addition, Castle erroneously thought that selection had somehow modified the Mendelian factor for the hooded pattern (for a discussion see Provine, 1971:113–114, 126–128).

William Castle initiated another series of selective breeding experiments to change the size of the dorsal stripe in hooded rats in 1907. These artificial selection experiments covered many generations of selection and involved breeding approximately 50,000 rats by the time the experiments were terminated in 1916 (Castle,

1911, Castle and Phillips, 1914, Castle and Wright, 1916, Castle, 1919). By 1911 these artificial selection experiments had produced evidence contradicting the contentions of the pure-line theorists concerning the effectiveness of selection in cross-breeding populations. Castle's successful artificial selection experiments provided strong evidence for three ideas that had been vigorously denied by antiselectionist Mendelians: first, that at least some continous variation had a genetic basis; second, that continuous variation exhibited a response to selection; and third, that selection could push variation beyond the limits of the pure-lines that were supposed to exist in populations (Mayr, 1973).

The statistical analyses of four experiments each involving 11 generations of selection of maize plants that F. M. Surface communicated to the Fourth Congress of Genetics in 1912 (Paper 23) also provided evidence for the effectiveness of selection operating on small variations. It is interesting to note that the American plant breeder Edward East (1910:198) had contended that after 12 generations, these maize selection experiments had "given complete corroboration of Johannsen's conclusions on pure lines."

These four maize selection experiments that involve selection from the maternal side only have been performed annually since 1896 at the Illinois Agricultural Experiment Station under the successive direction of C. G. Hopkins, L. H. Smith, C. M. Woodworth, and E. R. Leng. After 70 generations of selection the means of the selected characters (high protein, low protein, high oil, and low oil concentrations) in these four selection experiments were 215, 23, 341, and 14%, respectively, of the means (10.9% protein, 4.7% oil) of the maize kernals from the original population (Dudley, Lambert, and Alexander, 1974). Figures 1 and 2 taken from Dudley (1977) record the changes that have occurred in oil and protein concentrations each generation during the first 76 generations of the Illinois maize selection experiments. These four maize selection experiments are still being performed annually, and selection is still bringing about change in the selected populations (Dudley, 1977). These Illinois Agricultural Experiment Station selection experiments have been critically reviewed by "Student" (1933), Sprague (1955), and Wright (1977:186–191).

A. H. Sturtevant (1918) reported the results of artificial selection experiments for high and low bristle number in *Drosophila melanogaster.* Sturtevant concluded that selection can produce significant change in sexually reproducing populations, that selection produces its effects chiefly by isolating and accumulating numerous genetic differences already present in a population

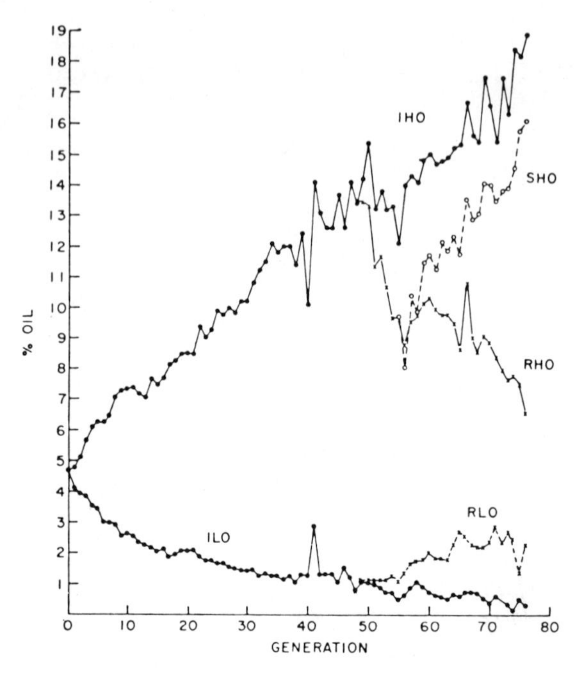

Figure 1. Mean percent oil for Illinois high oil (IHO), low oil (ILO), reverse high oil (RHO), reverse low oil (RLO), and switchback high oil (SHO) of maize seeds plotted against generations of selection. (From J. Dudley, 76 Generations of Selection for Oil and Protein Percentage in Maize, *Int. Conf. Quant. Genet., 1976, Proc.*, Ames, Iowa: Iowa State University Press, 1977, p. 462; copyright © 1977 by The Iowa State University Press.)

and that genes are relatively stable, not being contaminated in the heterozygote (contrary to the contention of William Castle and others; see Provine, 1971), and mutating very slowly.

The controversy as to whether evolution proceeds gradually or by jumps was resolved by artificial selection experiments and breeding experiments that involved quantitative traits. Small (continuous) variations that were heritable and that could be accumulated by selection were shown to exist. Evolution was experimentally demonstrated to occur as the result of the selective accumulation of small (continuous) variations as well as by the accumulation of large (discontinuous) variations.

The rapid development of genetic theories based on Mendelian patterns of inheritance led to a better understanding of artificial selection and evolution (Lush, 1950). Mendel's principles of segregation and recombination of genes demonstrated why

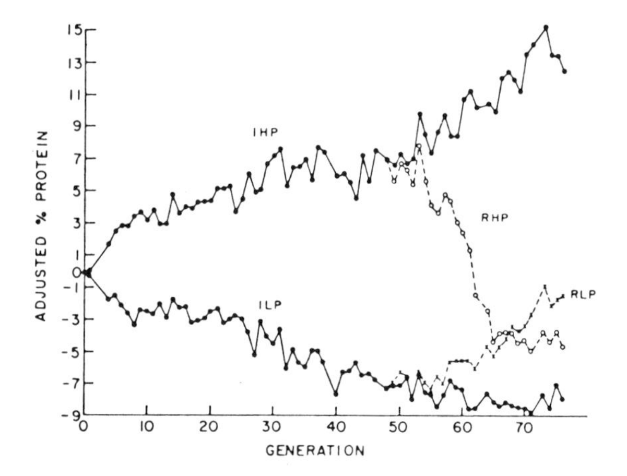

Figure 2. Mean adjusted percent protein for Illinois high protein (IHP), low protein (ILP), reverse high protein (RHP), and reverse low protein (RLP), of maize seeds plotted against generations of selection. (From J. Dudley, 76 Generations of Selection for Oil and Protein Percentage in Maize, *Int. Conf. Quant. Genet., 1976, Proc.*, Ames, Iowa: Iowa State University Press, 1977, p. 463.; copyright © 1977 by The Iowa State University Press.)

individuals with identical pedigrees could differ genetically. Second, Mendelian theory helped explain how an organism's genetic transmitting ability (its genotype) could differ from the expression of its genes (its phenotype). These two advances in genetics solved the mystery of atavism (reversion) and provided the theoretical framework for a more rational use of pedigrees and of progeny tests and sib tests in selecting breeding stock for practical and scientific artificial selection experiments (Lush 1950).

The effectiveness of artificial selection in bringing about evolutionary change has been demonstrated by numerous selection experiments performed during this century. After reviewing many of these artificial selection experiments, Sewall Wright (1977:559) wrote, "The most important conclusion is the practically invariable success of directional selection in bringing about drastic changes even from the narrowest foundation stocks. . . ."

18

Reprinted from pages 172–215 of *Selected Readings in Biology for Natural Sciences 3*, University of Chicago Press, Chicago, 1955

CONCERNING HEREDITY IN POPULATIONS AND IN PURE LINES*

By W. Johannsen

A Contribution to the Elucidation of Current Problems of Selection

Introduction: Aim of the Investigation

In no part of biology does the unity of life appear more clearly than in all the questions concerning fertilization and heredity. The most eminent investigators of these questions regard even the most dissimilar organisms from the same point of view; and in recent years, with increasing interest in questions of heredity, it is becoming more and more beautifully evident that the same general results will be obtained, through thorough-going studies, whether these studies concern "the crown of creation" or dogs and horses, hens and mice, moths and plant lice, or finally evening primroses and corn poppies, beans and peas, barley and oats.

This consideration gives me the courage—and also the right—to give a general title to the present work, although we are here concerned exclusively with investigations whose objects belong to the plant kingdom.

The present theory of inheritance, as it has developed particularly under the influence of the pioneering works of Francis Galton, is concerned less with the single individual than with jointly-ordered brother-and-sister lines and larger populations, and handles such groups as units.[1] Through statistical methods, partly also through

---

* A translation by Harold Gall and Elga Putschar of Ueber Erblichkeit in Populationen und in reinen Linien (Jena: Gustav Fischer, 1903). Translation by permission of the publisher.

1. Galton, Natural Inheritance, London, 1889, p. 35. See also Pearson's characteristic expression in the paper "Regression, Heredity, and Panmixia" (Mathematical Contributions to the Theory of Evolution III; Philosophical Transactions of the Royal Society, Vol. CLXXXVII, London, 1897, p. 255).

experiments, Galton, and after him particularly Karl Pearson, has discovered a series of regularities and has more precisely defined what appears, with more or less distinctness and purity, to have validity almost everywhere where it is a matter of inheritance within populations of pure types. The special relations of hybrids, it is true, do not generally comply with the Galton-Pearsonian rules—a question which is now causing lively discussion.[1]

Of the rules just named, the so-called throw-back or regression law is the most important. This law, in the particularly significant relation between parents and off-spring, expresses the fact that children (completely developed, mature), generally depart from the type of the given population in the same direction as the parents but in lesser degree. The children of deviating parents stand, on the average, nearer the so-called "type" of the population in question than do the parents.

With this chosen, much more than given, starting-point: the consideration of the population under investigation as a unit—be it a society, or a stock of animals or of plants of a given species (or race)—Pearson and his school in particular have given a mathematical treatment to the regression law, which, as far as I can judge and follow it, gives the impression of a superior statistical competence. And Pearson deserves above all the thanks of biologists; after Galton, this English mathematician has been to us perhaps the outstanding teacher of the exact treatment of experimental data; and his precise definitions of several of those concepts with which genetic theory operates are remarkably clarifying. In the theory of heredity there was—and still is—much too much loose talk.

As for actualities, it is sufficient for orientation to emphasize that Galton already believed not only that the relation between parents and offspring could be numerically expressed, but also that the average effect of individual parents, as well as of the grandparents and

1. Concerning hybrids see the works of de Vries (Die Mutationstheorie, Vol. II, 1903) as well as of Correns, Tschermak among others afterward. Particularly interesting here is Bateson's publication: Mendel's Principles of Heredity, Cambridge 1902. Discussion of the present question is conducted especially in the publications of the Royal Society, as well as in the journal Biometrika, mentioned here on p. 6.

still older generations, could be computed in such a way that the character of the progeny (their deviation from the mean character of the given population) seen on the average, is a definite function of the character of the different ancestral generations. Galton has gained his interesting results especially through studies of height, among other numerically expressible characteristics of the English people, as well as through investigations of the color relations of dachshunds and through cultivation tests with *Lathyrus odoratus*, of which the seed-size was examined. Pearson has significantly elaborated these questions, not only through exhaustive mathematical studies, but also by providing a new series of observations.

Pearson has expressed himself in the strongest terms upon the importance of the Galtonian regression law; thus he says in his brilliantly written *Grammar of Science*, p. 479:

> If Darwinism be the true view of evolution, i.e., if we are to describe evolution by natural selection combined with heredity, then the law which gives us definitely and concisely the type of the offspring in terms of the ancestral peculiarities is at once the foundation-stone of biology and the basis upon which heredity becomes an exact branch of science.

Readers who wish more precise information on the known regularities in populations I can refer to Pearson's own works.[1] It will often be rather difficult for the biologist to follow Pearson's arguments because of their rigorous mathematical form and because he generally seems to assume a special basic knowledge on the part of his readers.

It is now easily understandable that a more thorough-going biological study of hereditary ratios can by no means be satisfied with such essentially statistical investigations. A race, a people, a stock of any kind—let us call it from now on a "population"—is, biologically viewed, not always to be comprehended as a unit, nor need it be one

1. Besides the work named on p. 1 one finds a long series of Pearson's papers under the collective title: "Mathematical Contributions to the Theory of Evolution" in *Transactions of the Royal Society of London* and also in *Proceedings of the Royal Society*. Pearson gives a complete presentation of his general views in *The Grammar of Science*, 2nd edition, London, 1900 (the chapters "Life" and "Evolution," l.c.p. 328 to 503). There one finds further references to the writings of the authors named.

even though for the separate characteristics an average, supposedly "typical," value can be found around which the variations of all individuals under consideration fluctuate in the well-known fashion expressed by the "ideal variation curve" (the curve of the exponential law of error).[1] Such a population can—as will also be made clear in this work—include different independent types, very strongly varying from one another, a fact which need by no means be apparent from direct inspection of the empirical curve or table.[2]

Therefore, before one treats a population as a unit, one needs to analyze it biologically, in order to understand its elements, i.e., to take into account independent types already existing in the population. Only then is one able to decide whether and to what extent a uniform treatment is permissible. It is in reality such an analysis of a given race which Louis Leveque de Vilmorin had in mind when, supported by his comprehensive knowledge, he advanced his "isolation principle,"[3] a principle that can best be denoted as the principle of the individual estimation of progeny. According to this principle, one must, by means of isolation—keeping separate the seeds of every single mother plant—determine whether the "force of heredity"(force héréditaire), as Vilmorin expresses it, is great or small. We touch here on the notion "individual potency," so much disputed—especially earlier in the theory of horse and cattle-breeding; a notion which I would rather not bring into this discussion.

It is this same important principle which Hj. Nilsson, the distinguished director of the Swedish experimental station in Svalöf, has

1. Concerning the curves of variations and their construction I can refer to the excellent little introduction of Davenport: Statistical methods with special reference to biological variations, New York, 1899. Also, in Duncker's Die Methode der Variationsstatistik one will find details of variation curves. That the concept "type" or "typical value," as statisticians mostly define it (see Lexis, Abhandlungen zur Theorie der Bevölkerungs- und Moralstatistik, Jena 1903, p. 101) does not have to agree with the concept "type" of genetic theory—as this idea ought to be taken—will appear from this work.
2. There is an illustration—but only one illustration—of this in Pearson, Grammar of Science, p. 385, Fig. 26, where he discusses variations in the race and within the individual.
3. The works of Vilmorin appertaining to this are collected in the publication, Notices zur l'amélioration des plantes par les semis. Paris 1886.

already supported, for more than ten years, in directing his attention particularly to characters of the sort defined as "botanical" (i.e., morphological).[1] Nilsson inclines to the opinion that only by means of "botanical" signs can one control the purity of the pedigree-stock. A discussion could well be carried on about this, especially in view of the results of modern statistics of variation;[2] but such a discussion is unnecessary here. Nilsson and his school have none the less the quite indisputable merit of having demonstrated that supposedly pure races of cultivated plants "can, even when they appear completely homogeneous, consist of several independent forms with different properties, and so also with a different degree of winter-hardiness"[3]—one sees in these remarks a physiological character mentioned. In actuality physiological characters are just as important and valuable as morphological characters; to be sure it is usually more readily feasible to check the morphological characters if the question is one of constancy or of variability. The evaluation of physiological characters nearly always presupposes special lines of investigation, since the variability fluctuates strongly; however this does not touch the heart of the matter.

As far as I have been able to follow the publications of the Svalöf Institute, their findings add up to the viewpoint that every independent form type is constant; moreover it is to be understood that automatic, continuing, unilateral selection of variants does not lead to a gradual shifting of the type but new types arise through hybridization, or—as is most often the case—spontaneously, i.e., through mutation in de Vries' sense. Already in the year 1892 Nilsson had expressed this idea clearly and plainly,[4] an idea which has undoubtedly been the foundation of all later work at Svalöf.

---

1. There is a long series of detailed articles by Hj. Nilsson and his co-workers (Tedin, Ehle and others) in Sveriges Utsädesförenings Tidskrift in the years 1892 to the present. A summary in Vol. IX, 1899.
2. See e.g., Fr. Heincke's famous work Naturgeschichte des Herings, Part I. Die Lokalformen und die Wanderungen u. s. w. Berlin 1898.
3. Sveriges Utsädesförenings Tidskrift, 1901, p. 155.
4. Almänna Svenska Utsädesföreningens Tidskrift, Årgång II, 1892. no. 4, p. 131.

This conception, for which, it is true, reasons have not yet been more precisely given at Svalöf, corresponds completely, so to speak, with Hugo de Vries' mutation theory. I shall not have to decide here whether the hitherto published cultivation tests in Svalöf actually demonstrate the occurrence of mutations, as do in my opinion the famous cultures of de Vries. If I emphasize the Svalöf Institute, it is especially because they have worked there during a longer series of years with more rigorous application of the above-named Vilmorin principle and consequently have also attained results which—quite apart from their otherwise great interest—must be a very valuable support for my conception, to be mentioned below, concerning the nature of the Galton-Pearsonian laws. I am, therefore, also firmly convinced that my arguments will obtain the full agreement of the Svalöf investigators. Their findings and mine supplement one another admirably.

In Svalöf, as said above, it is particularly the "botanical" characters which are investigated; my investigations concern, first of all, "non-botanical" characters and I have judged them by means of, in part, very great numerical data. By this very fact, I am able to bring my results into most exact rapport with the Galton-Pearsonian investigations; it is just this which cannot be done with the beautiful Svalöf experiments, and, as is well-known, also did not and cannot happen in a satisfactory way with the mutations of de Vries.

My task here then is this: by means of investigations after Galton's and Pearson's methods, but using Vilmorin's principle, to test the range of the law of regression—first of all only in the relations between parents and offspring. By this it will be possible to show whether a real difference exists between mutation and fluctuating variability. While such eminent investigators as Hugo de Vries and W. Bateson[1]—to name only two of the leading workers, a botanist and a zoologist—definitely hold fast to such a distinction and seek to point it up as sharply as possible, Weldon, who here can be named as an excellent representative of the "biometrical" trend,[2] is by no

1. Bateson's chief work in this connection is Materials for the Study of Variations, London, 1894.
2. See Weldon's review of de Vries' Mutationstheorie in Biometrika, a journal for the statistical study of biological prob-

means willing to acknowledge this distinction. Weldon speaks of the efforts to define the difference as follows:

> These attempts appear always to rest upon a fancied relation between the phenomenon of "regression" and the stability of specific mean character through a series of generations which a little knowledge of the statistical theory of regression will show to be wholly imaginary.[1]

These are clear words and they contain an unfortunately not unjustified attack on the occasionally inexact experimental methods of certain modern biologists. But statistical theory alone surely cannot clarify the fundamental problems of biology!

In a population where the choice of mates is more or less free—as in human society—or where entirely pure random mating or fertilization occurs, as is the case with many animals and with cross-pollinated plants, a more precise analysis of the narrowest species, or race, in question, will be impossible, or carried out only through difficult and long-continued isolation experiments. In the case of plants with more distinct and reliable self-pollination the fluctuating variability is scarcely less than where cross-pollination prevails;[2] however one has the great advantage of being able to work freely and easily with "pure lines." By a pure line, I denote individuals which descend from a single self-fertilized individual. It is clear that a population of exclusively self-fertilizing individuals consists of absolutely pure lines, lines whose individuals, although mixed in nature, do not affect one another through crossing. And it appears to me that the relationships of pure lines must be the true basis of the theory of heredity, even though one cannot deal with pure lines in most populations, especially in human society. In any event, the behavior of pure lines is the simplest case, whose study must be carried out

---

lems, edited in consultation with Francis Galton by W. R. F. Weldon, Karl Pearson, and C. B. Davenport, Vol. I, Part III, April 1902, p. 365. I give the complete title of this new journal in order to show what outstanding investigators lead or rather support the "biometrical," i.e., mathematical, trend in genetic theory.

1. Biometrika l. c. p. 374.
2. I give my opinion here as an estimate; see besides Biometrika, Vol. I, Part I, p. 129 ff. (plant lice), and Pearson, Grammar of Science, p. 473.

before one can have full understanding of the more complicated cases, in which the more or less divergent lines cross reciprocally.

After these introductory remarks, I can proceed to examination of the regression law. This is easiest if Galton's own figures are taken as the starting point. I choose figures which concern the stature of part of the English population.

In Galton's data,[1] the ratio of the average height of the men to that of the women is 108:100. In calculating the height of a parental couple the height measurement of the woman will hence be multiplied by 108; thereby a legitimate correction is introduced, by means of which all height data are expressed in male heights. If now the parental pair are classified according to their average (male) height, one gets the following picture of the relation between parents and (adult) children, whose height is given of course in the same manner. The data are in English inches. The few cases in which the heights of the mid-parents* lie over 73 or under 64 inches are here excluded from the table.

Heights of mid-parents	64.5	65.5	66.5	67.5	68.5	69.5	70.5	71.5	72.5
Average height of children	65.8	66.7	67.2	67.6	68.2	68.9	69.5	69.9	72.2

If the average in both generations alike is set at 100, then one gets the following relative figures.

Parents	94	95.5	97	98.5	100	101.5	103	104.5	106
Children	96	97.5	98.5	99	100	101	101.5	102	105.5

This summary shows that parents who deviate from the average height of the population beget children who, seen on the average, deviate in the same direction but in lesser degree. The average deviation of the children was here about two-thirds of the parental deviation.

1. Natural Inheritance, p. 208. The modifications of the Galtonian tables, which Pearson has given in his above mentioned paper concerning "Regression, Heredity and Panmixia," have no special interest in this connection. Here Galton's original reports are the best suited to illustrate the matter.

* [Galton's term for the average of the male and transmuted female heights. Trans.]

Galton has found in a similar way that progeny (seeds) of selected small or large seeds of Lathyrus odoratus became smaller or larger, respectively, than the average seed-size for the species in the year in question. His figures relative to this[1] are not so regular as in the human data; but here also a similar "throw-back" is seen: the average deviation of the children was here perhaps only one-third the deviation of the parents.

By continued selection one should expect therefore that the character of a population could be shifted in one or the other direction,[2] although the throw-back, the regression, must act as an inhibiting or at least a delaying factor. In regard to this question, there are two contradictory opinions. Thus Hugo de Vries believes that the displacement can go only to a certain relatively narrow limit, and that the maximum of the displacement will be reached after few generations, when only one character is in question. Further selection will then cause no further shifting, but only maintain the level reached. De Vries leans here partly on his own investigations, partly on the great experimental data from the field which he himself reviewed in a critical manner.[3] Of pure lines in my sense there is here however, as far as I can see now, no mention.

In sharp contrast to this conception stands that of the biometric school, which maintains that continued selection can produce a continual shifting of the average character. Naturally there are here also (tacitly) limits assumed—a beet-root cannot for example consist of pure sugar—however the limits are wider and attained much later than according to Hugo de Vries' notion. At first sight, the biometrical notion is by no means absurd. I see no experimental evidence for its correctness; it leans largely on the "statistical theory of regression," to use the Weldonian expression (see p. 7). It appears to me, however, that they want to ascribe a little too much

1. Natural Inheritance, p. 226.
2. Thus Ammon imagines a shifting easily effected through selection: see his publication "Der Abänderungsspielraun" (Naturwiss. Wochenschrift, 1896, No. 12-14. The whole Ammonian conception is based on thought-experiments, interesting in themselves, which we need not deal with here.
3. Die Mutationstheorie, Vol. 1, p. 83 ff.

importance to statistical theory.

Over a number of years I have worked with questions of heredity, partly with populations, partly with pure lines. My investigations relate intentionally only to self-pollinating plants: two-rowed barley (4 cultivated strains of f. nutans and 2 hybrid strains), beans (Phaseolus vulgaris; 2 different bush-beans), and peas (Pisum sativum). The investigations have concerned diverse special questions and are only partly published.[1] In the present paper only the relationship of my results to the Galtonian regression law (between parents and children) will be considered. It is hoped that in later publications a series of particular results can be given.

What I want to bring to the clarification of the law in question concerns three characters of quite different nature. The first is a quantity, namely seed-size, i.e., the weight of the beans. This character corresponds completely to that which Galton examined in the case of Lathyrus odoratus, especially since this plant is also self-pollinated. The seed-size can be very dependent on external circumstances; especially do the meteorological conditions of the particular year have a great influence. With culture after simultaneous sowing in the same well-cultivated beds and with equal, abundant soil-room, the comparison within the given year's set is fully justified. In any case the result agrees perfectly with the behavior of other characters.

The second character is the relative width of beans (the width expressed in per mille of the absolute length); accordingly a derived value, a relation between two quantities —I may perhaps say a kind of quality. This character is not dependent on the life-situation in different years in the same degree as the seed-size, although clearly an influence is seen, which must be taken into account.[2] The length and breadth can and must, of course, be examined; they behave in the same manner as the seed-weight. However, in order to have a char-

1. In the paper "Om Variabiliteten med saerligt Hensyn til Forholdet mellem Kornvaegt og Kvaelstof-Procent hos Byg" (Meddel: fra Carlsberg Laboratoriet, Vol. IV, p. 228-313, 1899, detailed French summary) I have reported on several questions of correlation. I must here strongly emphasize that the numerical treatment in this paper was very primitive.
2. The regression must of course be judged with the mean character of the progency concerned in the given year as starting-point. See Weldon's critical comments in Biometrika, I, p. 368.

acter of another nature than size, the relative width, that is to say the form, was chosen. The correlation between length and breadth is, moreover, the object of particular studies which are not yet completely finished.

The third character is the greater or lesser tendency of certain barley forms to be "jagged," i.e., to form ''blanks" in the ears, since an often significant number of ovaries do not set. Here we are dealing with an, in many cases, hereditary abnormality, and, what is more, considering a monoctyledonous plant. This character also agrees completely with the two preceeding.

This full agreement between three quite essentially different characters in different plant species, and in different races within the species, may certainly give my data a greater interest than the behavior of a character standing alone can claim.

FIRST RESEARCH SERIES

SEED-SIZE OF BEANS

The starting point of these investigations was a purchased lot, about 8 kg., of brown "Princess" beans, one of the oldest bush-beans among the many cultivated forms of Phaseolus vulgaris. These beans were harvested in the year 1900 on the island Fühnen, and were exceptionally fine and as uniform on the whole as can be expected here. The average seed-weight was 495 mg. determined with about 5,000 beans taken at random.

Now, in the spring of 1901, 100 seeds were chosen which as nearly as possible represented the average character of the lot in regard to length, breadth, and weight. Their average weight was about 500 mg.; they varied only from 470 to 530 mg. Further, the 25 smallest as well as the 25 largest were chosen out of the whole lot of about 16,000 beans. At the same time, all the seeds chosen for sowing with another purpose in mind (in regard to the relative width), were weighed, so that here also the sown seeds could be grouped according to weight classes. Each seed was sown in a definite numbered place in the experimental garden covered over with wire netting, each seed having abundant room at its disposal, about 80 sq. cm. For the present the question of inheritance of weight was a secondary concern; but the crop contributed to the clarification of this matter. The results are grouped, summarized, in the table below:

Seed weight (of the individual seeds) of brown beans, 1901

All weight data in milligrams. The division into classes, arranged from left to right and labeled above, indicates the limits of weights of the mother-seeds in question.

Classification of the mother-seeds:	250–350	350–450	450–550	550–650	650–750	750–900
Number of seeds sown	25	40	161	24	12	25
Average weight of the crop .	374	388	401	434	446	457
No. of seeds in the crop	606	914	4476	604	370	598
No. of seed-bearers	19	26	123	17	11	11

The table confirms the regression law in so far as it clearly appears that the offspring of deviating mother-seeds deviate, on the average, in the same direction but to lesser degree. Of course, for the offspring generation the particular so-called "type" of the progeny —here accordingly the average bean-weight of the offspring of average parents in the particular year, under the given circumstances— must be the starting-point for the more precise determination of the relation between the deviation of the mother-seeds and the harvested seeds. In this case I have no reason to make a more precise comparison of this kind. For this, it would be desirable also to weigh all individuals of the crop separately; this was done only with the offspring of the smallest and largest mother-seeds: in these seed-crops all single seeds were weighed separately and distributed into classes each with a range of 50 mg. (5 cg.).

Both these seed-crops, whose mother-seeds (1900) had an average weight of about 280 and about 820 mg., respectively, showed the following variation relationships:

1901 offspring of the small beans

Classification[1]	10–15	15–20	20–25	25–30	30–35	35–40	40–45	45–50	50–55	55–60	60–65
Number of seeds		8	18	71	156	172	127	35	15	3	6
"Theoretical" numbers .	1	5	25	74	139	164	122	57	17	3	

The average of these weighings is 368.4 mg., the standard deviation[2] $\sigma = \pm 71.9$ mg. From this, with the familiar formula of the exponential law of errors, the "theoretical" numbers of the lowest horizontal row are calculated. These calculated figures agree as well with the observed, on the whole, as was to be expected here (compare the agreement in the pure lines, illustrated by figures 3-5, pp. 25-26). After this it seemed wholly justifiable to regard the offspring of the small beans as belonging to a single weight-type, which a glance at the neighboring Fig. 1 would confirm.

1. Here 10, 15, 20 centigrams are written instead of 100, 150, etc., milligrams.
2. I permit myself here the designation standard-deviation, or to use the letter σ to express "the root of the average squared deviation." "Average deviation" has, in German, another significance. See Davenport, Statistical Methods, p. 15.

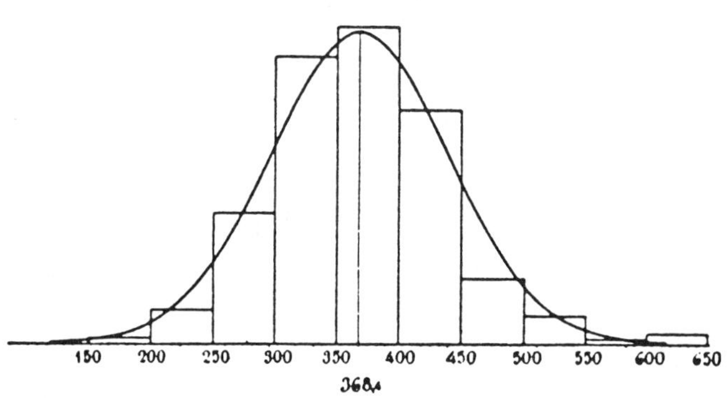

Fig. 1. The variation of seed-weight of 606 beans, 1901 offspring of the lowest weight-class of 1900. The curved line displays the "ideal" variation curve (the exponential curve of error). The foot of the vertical line, which bisects the area of the curve, tells the average of all variants. The area of the rectangle corresponds in the class in question to the observed number of individuals. The limits of the weight-classes are marked on the abscissa axis (base-line) measured in tenths of the standard-deviation. In that the class range is 50 mg. and the standard deviation $\sigma = \pm 71.9$ mg., the width of each range becomes $50 : 7.19$ = about 6.95 tenths of σ.

In this paper all the graphical presentations of the variation relationships are constructed using a fixed scheme as starting point. An ideal curve was constructed once and for all after the well-known formula of the theory of error and in arbitrary proportions of scale, such that the area of the curve corresponds to 10,000 units. The drawing was executed in zinc-etching and printed for my work. In such a scheme it is easy to mark in the rectangles (see the figures) according to previous conversion of the number of individuals to per mille, and with calculation of the curve ranges in tenths of the standard deviation. That is, the abscissa axis on the scheme, which is about three times bigger than the reproduction given here, is plainly divided into tenths of the standard deviation; there is also a scale on the scheme, to permit setting down the height of the rectangles. This scale is only of interest in carrying out the drawing and is not reproduced here. As for the rest see the illustration of Fig. 1. Although these offspring of the smaller beans thus show a very good, normal distribution of variants, they nevertheless consist of rather differently constituted "pure lines" (see p. 7). By considering each plant separately it appeared, directly after the harvest, that great differences occurred between the seeds of the different plants, although the mother-seeds were the same size. If the average weight of the seeds of each plant is calculated separately, then the 19 lines are distributed as follows:

Classification	300	350	400	450	500	550
No. of lines in question	5	7	6		1	

This behavior shows clearly what the situation is here: through selection of the very small seeds of the 1900 crop one has ended up with proportionately many which belong to the small-seeded type (under 350 mg.) and only one which belongs to the large-seeded type (over 450 mg., in the particular year). Through this the regression of the progeny as a whole will be easily understood.

And the same also for the offspring of the very large seeds:

1901 offspring of the very large beans

Classification:[1]	10	15	20	25	30	35	40	45	50	55	60	65	70	75	80
No. of seeds in question . .			5	18	46	144	127	70	70	63	28	15	8	4	
Theoretical numbers . . .	1	3	11	26	54	85	109	111	91	59	30	13	4	1	

The average of these single weighings was 454.4 mg. and the standard deviation $\sigma = \pm 104.2$ mg. The theoretical numbers calculated from these agree poorly with the observed; it is not necessary to illustrate this by graphical presentation. It is probably caused in part by the fact that here fewer lines were present, to wit, only 11. These were also very diverse; classified as in the case of the small beans above, we see the following distribution of lines:

Classification:	350	400	450	500	550	600
No. of lines in question	4	2		2	3	

1. See note 1, p. 13.

This distribution shows clearly that by selection proportionately many seeds belonging to large-seeded types, and not any belonging to small-seeded types—under 350 mg.—are obtained.

The importance of the Vilmorin principle will at once be understood by inspection of these numbers, and at the same time a glance at Fig. 1 will show that the rather good variation curve by no means states whether there is one or more than one type. It is clear that an irregularity in the distribution of individuals, as was the case in the variation of the offspring of larger beans, points to differences of type. We will return to these questions later.

If thus, seen in general, the regression law of the previously reported experiments also is confirmed here by means of the average numbers, then it becomes immediately obvious that the value of these averages as expression of one type each is highly questionable. And we must ask therefore whether or not, through selection of plus- and minus-variants within the pure lines, a type displacement, or rather a Galtonian regression, could be achieved.

In order to clarify this fundamentally important question a number of the most fruitful lines of 1901 (seeds of one seed-bearer each) was used. Each line was thus descendant of one well-characterized single bean from the 1900 crop. These lines were now sown in the spring of 1902 and harvested in fall of the same year. Of the 1902 crop, however, not merely these especially chosen lines (designated with the letters B, G, H, J, N, and O) were used; but all lines for which individual seed-weighings were at hand were employed together toward clarifying the matter. In this way, on the whole, data concerning 19 lines were obtained. Each line thus was derived from one bean of the 1900 crop and the total number of individually weighed beans of the 1902 crop amounted to 5,494. The weighings were carried out by means of a balance especially constructed for the task.[1]

Within each line the seeds of the 1901 crop (the "mother-seeds") were grouped into weight-classes, with the class-range generally used here of 50 mg. The progeny seeds of 1902 of these weight-classes—each seed-bearer considered separately—were in turn

1. The balance was manufactured by the firm Levring & Larsen, Copenhagen V.

divided in similar manner into weight-classes.[1] It would of course take too much space to give here all the numerical details obtained. It will be enough first of all to characterize each line separately,[2] and to examine within the separate lines the seed-progeny (1902) of each mother-seed weight-class (1901). In this way a sufficiently detailed account of the behavior of the material will be given.

In the following grouped tables of lines, all weight data are expressed in milligrams.

Line A. Weight of the grandmother bean (1900): ca. 800.
Average weight of the mother-seeds (1901): ca. 600.

Weight classes of mother-seeds[3] (1901)	Character of the offspring (1902)		
	Average Weight	Number of Seeds	Standard Deviation[4] $\pm\sigma$
550-600	605	15	126.8
600-650	642	39	107.9
650-700	635	45	105.8
700-750	661	46	112.4
The entire line	641.9	145	109.5

1. The harvested plants are first placed in numbered paper bags and kept about two months in a cool room. The seeds of each individual plant are then placed in a suitably large cork-stoppered specimen tube. By this no disturbance through drying, etc., can occur. Generally the individual seeds are also numbered by means of India ink, i.e., they receive, in the case of each plant, a running number, beginning with 1.
2. Details concerning the variation within each line on p. 22.
3. Where not all seeds of 1901 are sown, the average weight of all sibling seeds is of course still given.
4. The "probable error" of the mean is given in the table, pp. 28-29. In calculating the standard deviation I have here divided by (n−1) instead of n as is permissible in the case of large numbers.

Line B. Weight of the grandmother bean (1900): 950.
Average weight of the mother-seeds (1901): 520.

Weight classes of mother-seeds (1901)	Character of the offspring (1902)		
	Average Weight	Number of Seeds	Standard Deviation ±σ
350-400	572	86	85.1
450-500	535	118	89.7
500-550	570	77	105.4
550-600	565	72	98.4
600-650	566	48	65.7
650-700	555	74	98.4
The entire line	557.9	475	93.0

Line C. Weight of the grandmother bean (1900): ca. 545.
Average weight of the mother-seeds (1901): 570.

500-550	564	144	78.9
600-650	566	40	65.7
650-700	544	98	77.1
The entire line	554.3	282	76.4

Line D. Weight of the grandmother bean (1900): 475.
Average weight of the mother-seeds (1901): ca. 600.

500-550	542	32	60.9
550-600	536	163	92.0
650-700	566	112	77.2
The entire line	547.6	307	84.1

Line E. Weight of the grandmother bean (1900): ca. 540.
Average weight of the mother-seeds (1901): 512.

400-450	528	107	87.7
500-550	492	29	47.5
650-700	502	119	66.5
The entire line	511.9	255	75.9

Line F. Weight of the grandmother bean (1900): ca. 280.
Average weight of the mother-seeds (1901): 395.

Weight classes of mother-seeds (1901)	Character of the offspring (1902)		
	Average Weight	Number of Seeds	Standard Deviation $\pm\sigma$
300-350	485	117	75.2
450-500	479	124	77.5
The entire line	481.8	241	76.3

Line G. Weight of the grandmother bean (1900): ca. 800.
Average weight of the mother-seeds (1901): 400.

250-300	421	28	107.1
350-400	490	105	93.8
400-450	459	307	64.1
450-500	469	93	86.1
The entire line	465.1	533	78.8

Line H. Weight of the grandmother bean (1900): 433.
Average weight of the mother-seeds (1901): 380.

300-350	452	114	83.1
350-400	469	81	59.2
400-450	445	136	69.7
450-500	462	87	59.5
The entire line	455.3	418	70.1

Line J. Weight of the grandmother bean (1900): ca. 800.
Average weight of the mother-seeds (1901): 400.

250-300	445	40	84.0
300-350	498	53	64.8
350-400	453	164	69.9
400-450	447	155	65.5
450-500	434	103	74.9
500-550	468	102	79.1
550-600	458	95	79.9
The entire line	454.4	712	74.0

Line K. Weight of the grandmother bean (1900): 270.
Average weight of the mother-seeds (1901): 510; very unlike in **weight.**

Weight classes of mother-seeds (1901)	Character of the offspring (1902)		
	Average Weight	Number of Seeds	Standard Deviation $\pm\sigma$
200-250	469	18	59.4
450-500	446	131	68.1
600-650	450	39	63.7
The entire line	449.5	188	66.8

Line L. Weight of the grandmother bean (1900): ca. 380.
Average weight of the mother-seeds (1901): 360.

Weight classes of mother-seeds (1901)	Average Weight	Number of Seeds	Standard Deviation $\pm\sigma$
300-350	459	147	74.6
350-400	441	90	55.2
450-500	410	36	70.2
The entire line	446.2	273	69.4

Line M. Weight of the grandmother bean (1900): ca. 270.
Average weight of the mother-seeds (1901): 340.

Weight classes of mother-seeds (1901)	Average Weight	Number of Seeds	Standard Deviation $\pm\sigma$
200-250	440	78	72.4
350-400	424	217	71.4
The entire line	428.4	295	71.8

Line N. Weight of the grandmother bean (1900): 600.
Average weight of the mother-seeds (1901): 312.

Weight classes of mother-seeds (1901)	Average Weight	Number of Seeds	Standard Deviation $\pm\sigma$
150-200	410	54	55.0
250-300	422	111	76.7
300-350	389	92	89.4
350-400	408	100	78.0
The entire line	407.8	357	78.5

Line O. Weight of the grandmother bean (1900): ca. 270.
Average weight of the mother-seeds (1901): 310.

Weight classes of mother-seeds (1901)	Average Weight	Number of Seeds	Standard Deviation $\pm\sigma$
250-300	358	72	55.7
350-400	348	147	66.4
The entire line	351.3	219	65.0

Line P. Weight of the grandmother bean (1900): ca. 500.
Average weight of the mother-seeds (1901): 390.

Weight classes of mother-seeds (1901)	Character of the offspring (1902)		
	Average Weight	Number of Seeds	Standard Deviation $\pm\sigma$
250-300	454	21	68.2
400-450	469	51	79.4
500-600	428	34	67.5
The entire line	452.8	106	75

Line Q. Weight of the grandmother bean (1900): ca. 500.
Average weight of the mother-seeds (1901): 440.

Weight classes of mother-seeds (1901)	Average Weight	Number of Seeds	Standard Deviation $\pm\sigma$
200-250	459	16	81.7
350-400	495	262	67.4
600-650	482	27	70.7
The entire line	492.0	305	69.0

Line R. Weight of the grandmother bean (1900): ca. 500.
Average weight of the mother-seeds (1901): 410.

Weight classes of mother-seeds (1901)	Average Weight	Number of Seeds	Standard Deviation $\pm\sigma$
200-250	496	14	42.7
500-550	451	42	57.6
600-650	440	27	80.1
The entire line	455.1	83	66.0

Line S. Weight of the grandmother bean (1900): ca. 500.
Average weight of the mother-seeds (1901): 405.

Weight classes of mother-seeds (1901)	Average Weight	Number of Seeds	Standard Deviation $\pm\sigma$
300-350	490	20	55.0
350-400	491	119	68.9
500-550	475	20	102.6
The entire line	488.8	159	72.5

Line T. Weight of the grandmother bean (1900): ca. 500.
Average weight of the mother-seeds (1901): 395.

Weight classes of mother-seeds (1901)	Average Weight	Number of Seeds	Standard Deviation $\pm\sigma$
250-300	535	20	48.7
350-400	508	111	62.8
550-600	425	10	47.1
The entire line	506.2	141	64.3

To throw light on the variation of weight within the pure lines and in the entire material, individual determinations, arranged in the usual weight-classes, are collectively presented in the following tabular summary:

SUMMARY TABLE I

Weight relations of brown beans. Crop of 1902

The variation within pure lines and in the collected material.

The figures indicate the number of seeds in the weight classes concerned. Underlined type indicates the individuals who stand nearest the average weight of the line in question.

Classification:[1]	10	15	20	25	30	35	40	45	50	55	60	65	70	75	80	85	90 cg. Total
Line A	..	..	..	..	..	2	5	9	14	21	22	24	23	17	6	2	145
— B	..	..	..	1	6	19	32	66	88	100	90	50	19	1	3	..	475
— C	..	..	..	..	..	5	14	50	76	58	44	29	5	1	..	..	282
— D	..	..	..	5	2	9	21	38	68	77	62	22	3	..	..	..	307
— E	..	..	..	4	1	12	29	62	65	57	19	6	..	..	..	..	255
— F	..	..	..	2	8	21	46	74	46	28	14	1	1	..	..	..	241
— G	..	..	3	9	28	51	111	174	101	44	6	..	1	5	..	..	533
— H	..	..	1	6	20	60	106	114	75	33	3	..	..	..	..	..	418
— J	..	1	2	14	38	104	172	179	140	53	9	..	..	..	..	..	712
— K	..	..	1	2	6	31	55	55	28	6	4	..	..	..	..	..	188
— L	..	..	1	5	15	37	88	76	33	13	4	1	..	..	..	..	273
— M	..	..	4	9	26	56	82	76	32	9	1	..	..	..	..	..	295
— N	1	3	11	22	29	72	120	69	23	5	2	..	..	..	..	..	357
— O	4	4	5	19	69	69	44	5	..	..	..	..	..	..	..	..	219
— P	..	..	..	3	1	18	35	27	13	3	4	2	..	..	..	..	106
— Q	..	..	1	2	7	16	44	93	80	52	10	..	..	..	..	..	305
— R	..	..	..	2	3	12	17	27	19	3	..	..	..	..	..	..	83
— S	..	..	1	2	3	8	27	47	37	30	4	..	..	..	..	..	159
— T	..	..	..	..	1	6	20	37	39	30	8	..	..	..	..	..	141
The collected material	5	8	30	107	263	608	1068	1278	977	622	306	135	52	24	9	2	5494
Theoretical figures	2	8	36	121	318	636	973	1136	1014	691	362	144	43	10	2	0	

1. For lack of space 10, 15, 20 centigrams are placed here instead of 100, 150, 200 milligrams.

The mean weights and standard deviations of separate lines are given in the individual tables pp. 17-21. The mean weight of the collected material is 478.9 mg.; the standard deviation $\sigma = \pm 95.3$ mg. From these the theoretical (law of error) figures are calculated, given at the bottom of Table 1. The agreement with the observed numbers is fairly good, and is by no means less than the agreement found between calculation and observation in the majority of the pure lines, considered separately. Figure 2 illustrates the variation in the collected material; the figure is constructed according to the same principles as Figure 1.

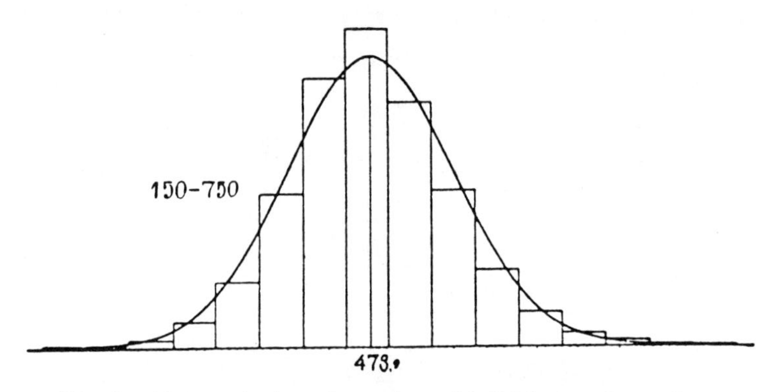

Fig. 2. The variation of weights of 5494 beans from the 1902 harvest, progeny of all weight classes of 1901. See the explanation to Fig. 1 and the description on p. 14.

The agreement which we observe here with the error law, or rather its curve, is, more closely considered, quite meaningless. Where there is a single type around which the variants group themselves symmetrically, it is obvious that one will get, with the conditions given here, a normal curve ("Galton curve"). But the existence of such a curve does not show at all whether the individuals belong to a single or to several types. The latter is the case here. The curve in Figure 2 —and something similar pertains, as has been said, to Figure 1—is not at all a homogeneous variation-curve. It expresses the totality of all variations of the different, more or less regularly varying, lines.

And within each line there is found again a double variation: the individual sibling series (the seeds of one seed-bearer each) can depart more or less from the average of the line, and within each sibling series the variation is often strongly marked.

Pearson[1] distinguishes between the variability of the race and of the individual. Within the race the different plants vary ("racial" variability) and in one and the same plant the different homologous organs again vary in respect to any one character; thus fruits, leaves, etc., are not alike in a given plant ("individual" variability).[2] One would like to consider the variability of the seeds of a single seed-bearer as example of the Pearsonian "individual" variability—which is scarcely permissible, as only the seed-coat belongs to the mother-plant—then one would be able to insert, in our present case, "lineal" between "racial" and "individual"; and it is easily seen that one can so group the whole series of variations within the species: "specific," "racial," "lineal," "fraternal" and "individual" ("partial," of De Vries) variability, with a somewhat decreasing range of variation on the whole.

Thereby, of course, the curves which express the higher categories of these variation-series lose nearly all meaning as expression of truly typical relationships; such curves then become hardly more than expression of—accidents. Fig. 2 is a good example thereof; it is the sum of the "accidentally" chosen lines, in which the individuals of the sibling series are grouped according to the law of large numbers; often, as here, symmetrically, frequently in more skewed distribution. This gives then absolutely no certain basis in regard to the question of the type (in the sense of the heredity) of the single individual. There is, no doubt, now and then some game or statistical sport carried on, perhaps as is at times practised by histologists with serial sections and staining methods, and by earlier agricultural chemists with ash analyses. Within the pure lines the variation curves[3] have nevertheless a quite different value than in larger populations. Therefore I give several curves here which illustrate the behavior of different lines.

1. Grammar of Science, p. 381; see p. 385, Fig. 36.
2. De Vries' Mutationstheorie, Vol. 1, p. 100, calls this "partial" variability (i.e. the variability of the parts of the individuals).
3. The polymodal (many-peaked) curves are particularly important here, as shall be presented in a later work.

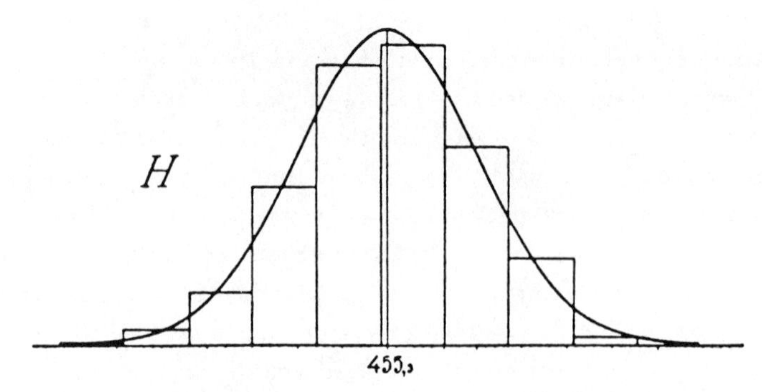

Fig. 3. The variation of seed-weight in line H (418 seeds).

In some lines, e.g., B, H, J, K, M, Q, the variation agrees very well with the ideal curve, illustrated by Figures 3 and 4.

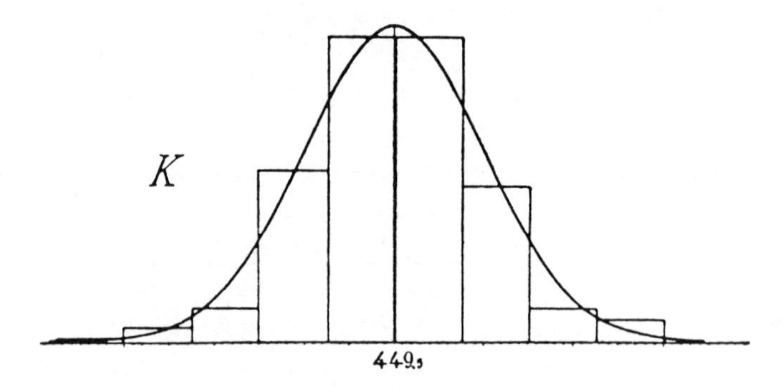

Fig. 4. The variation of seed-weight in line K (188 seeds).

In other cases, e.g., in lines A, F, G, and N, the deviation from the ideal curve is greater, as Fig. 5 illustrates. The particular seed-number of the line is not determinative here, as can be seen by comparison of the data in Figures 4 and 5.

By these examples one sees that very frequently too many individuals are found in that class (the "main class") in which the average value of the line lies (Fig. 5). Ludwig[1] has had, so far as I can see, similar curves, when he speaks of "hyperbinomial" curves and

1. Ludwig, Die pflanzlichen Variationskurven u. s. w. (Botanisches Centralblatt, Vol. LXXIII, 1898, p. 241 ff. See particularly p. 294.)

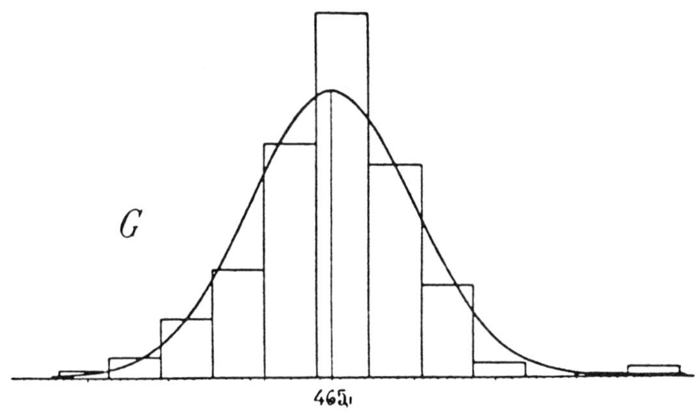

Fig. 5. The variation of seed-weight in line G (533 seeds). The "main-class" seems to be represented too numerously.

even seeks to explain these through the assumption of "invariants, whose larger or smaller number should raise the variation-polygon more or less markedly in the middle. The variation should thereby be steeper, the curve "better" than the normal curve.

It appears to me that this notion is very far-fetched. In my material the exaggerated peak is caused by the fact that relatively many individuals deviate very strongly to one or to both sides of the mean (compare line G and Fig. 5). It is obvious that the relatively large squared sum of these deviations will produce an increase of the standard deviation; because of this the number of individuals of the main class will necessarily be found too large. The deviating individuals act, if one wants to express it so, as "clumsy errors" in an observation series—and not invariants but rather relatively strong deviations produce the "hyperbinomiality." Especially where it is a matter of quantities (weight, length, breadth: in Ludwig's case the absolute number of different organs), which are for the most part directly influenced by purely local conditions, one has hyperbinomial curves, accordingly much more frequently with plants than with animals, without implying that one supposes a principal difference between the two kingdoms is found thereby—I remember having read something about it, but the work in question is lost to me. I believe that Ludwig's interpretations in any case need a closer examination; the eminent plant arithmetician will perhaps best be able to carry out this work himself; of course it is necessary for this purpose also to work with pure lines.

After these considerations of variability, the principal question, the effect of selection, must be cleared up. If as is mostly the case in biological-statistical investigation, one gives no consideration to pure lines but considers the material as a whole, one will undoubtedly find Galton's regression law confirmed here. Summary Table 2, p. 28 illustrates this.

I find no reason to determine more precisely according to what numerical proportion the regression established here occurs. I would only like to mention that my data, directly observed, are smoother than found by Galton for Lathyrus odoratus. These figures[1] give, namely, if the middle value—since I do not know the average—be set at 100, the series: 94, 98, 96, 100, 98, 106, 107. The adjusted figures of Galton give the series: 94, 96, 98, 100, 102, 104, 106—I nevertheless would not like to undertake such a rounding off of my data, because I regard it as superfluous here.

In order to examine regression in my material more closely, the average size of the mother-seeds of each class would have to be calculated to begin with. I also see no reason to group the data such that the classes of the offspring serve as the starting-point for the grouping of the mother-seeds, as Galton has done so interestingly with his data for human stature.[2] For me, it is enough to make clear that an average fractional regression in my material is clearly to be seen, which agrees thus far with Galton's statements.

The summary table shows how this result follows: in the lower-weight classes of the mother-seeds, lines were pre-eminently, but by no means exclusively, represented which were distinguished by low seed-weight; in the higher weight classes of the mother-seeds the contrary was the case, and in the middle class the majority of all the lines met. Thus, with selection of mother-seeds an only partial isolation of small-seeded, medium, and large-seeded lines was obtained; each selection-class represents a more or less variegated mixture of individuals of different lines. It is interesting to see how the variation proportions are arranged in the progeny of these impure selection-classes:

1. Natural Inheritance, p. 226, Table 2, next to the last column.
2. Ibid., p. 208.

SUMMARY TABLE 2

SEED WEIGHT OF BROWN BEANS, 1902

The data are in milligrams. The numbers after ± give the probable error (of the mean) of the determinations in question. The underlined figures specify the number of individual seeds which underlie each determination. The class divisions from left to right express the weight classes of the mother-seeds (1901) in question.

Classification of the mother-seeds	150–200	200–250	250–300	300–350	350–400	400–450
Number of seeds sown	6	14	32	55	150	78
Line A	....	....	....	....	....	....
— B	....	....	....	....	572±6 86	....
— C	....	....	....	....	....	....
— D	....	....	....	....	....	....
— E	....	....	....	....	....	528±6 107
— F	....	....	....	485±5 117	....	....
— G	....	....	421±14 28	....	490±6 105	459±2 307
— H	....	....	....	452±5 114	469±4 81	445±4 136
— J	....	....	445± 9 40	498±6 53	453±4 164	447±4 155
— K	....	469± 9 18	....	....	....	....
— L	....	....	....	459±4 147	441±4 90	....
— M	....	440± 6 78	....	....	424±3 217	....
— N	410±5 54	....	422± 4 111	389±6 92	408±5 100	....
— O	....	....	358± 4 72	....	348±4 147	....
— P	....	....	454±10 21	....	....	469±6 51
— Q	....	459±14 16	....	....	495±3 262	....
— R	....	496± 8 14	....	....	....	....
— S	....	....	....	490±8 20	491±4 119	....
— T	....	....	535± 7 20	....	508±4 111	....
Combination of all progeny without regard to the pure lines	410 54	453 126	419 292	456 543	460 1482	465 756
	440		443		461	
The average of all weights set alike at 100	92		93		96	

The following Summary Table 3 gives a further explanation. A variation curve for the total data was already given in Fig. 2 (see also Summary Table 1, p. 22). The next figures, 6 and 7, illustrate both

450–500	500–550	550–600	600–650	650–700	700–750	
70	52	42	31	38	6	Mean
....	....	605±22 15	642±12 39	635±11 45	661±11 46	642.0±6 145
535±6 118	570± 8 77	565± 8 72	566± 6 48	555± 8 74	....	558.2±3 475
....	564± 4 144	....	545± 7 40	544± 5 98	....	554.4±3 282
....	542± 7 32	536± 5 163	....	566± 5 112	....	547.6±3 307
....	492± 6 29	....	....	502± 4 119	....	511.8±3 255
479±5 124	....	....	....	....	....	481.9±3 241
469±6 93	....	....	....	....	....	464.9±2 533
462±4 87	....	....	....	....	....	455.1±2 418
434±5 103	468±5 102	458± 6 95	....	....	....	454.7±2 712
446±4 131	....	....	450± 7 39	....	....	449.0±3 188
410±8 36	....	....	....	....	....	446.6±3 273
....	....	....	....	....	....	428.2±3 295
....	....	....	....	....	....	407.8±3 357
....	....	....	....	....	....	351.2±3 219
....	....	428± 8 34	....	....	....	452.9±5 106
....	....	....	482± 9 27	....	....	492.0±3 305
....	451± 6 42	....	440±10 27	....	....	455.0±5 83
....	475±15 20	....	....	....	....	488.9±4 159
....	....	425±10 10	....	....	....	505.9±4 141
469 692	522 446	513 388	529 220	549 448	661 46	478.9 5494
490		519		560		
102		108		117		100

the cases where the variation in the classes agrees with the law of error. Fig. 8 on the other hand is the case where the agreement is least.

SUMMARY TABLE 3

WEIGHT RELATIONS OF BROWN BEANS. CROP OF 1902.

The variation of progeny of different weight-classes (double-classes of the mother-seeds, 1901).

The figures of the main table state the number of individuals in the column concerned.

Mother-seeds, double-classes	The variation of the progeny. Division into classes as in Summary Table 1																	Sum	Standard Deviation ±σ
		10–15	15–20	20–25	25–30	30–35	35–40	40–45	45–50	50–55	55–60	60–65	65–70	70–75	75–80	80–85	85–90		
Milligrams																		Number	in milligrams
150-250	..	..	..	1	3	12	29	61	38	25	11	..	..	..	..	..	..	180	69.6
250-350	..	..	2	13	37	58	133	189	195	115	71	20	2	..	..	..	..	835	87.0
350-450	..	5	6	11	36	139	278	498	584	372	213	69	20	4	3	..	..	2238	85.1
450-550	..	..	..	4	20	37	101	204	287	234	120	76	34	17	3	1	..	1138	91.8
550-650	..	..	..	1	9	14	51	79	103	127	102	66	34	12	6	5	..	609	102.5
650-750	..	..	..	..	2	3	16	37	71	104	105	75	45	19	12	3	2	494	97.1
The total material		5	8	30	107	263	608	1068	1278	977	622	306	135	52	24	9	2	5494	95.3

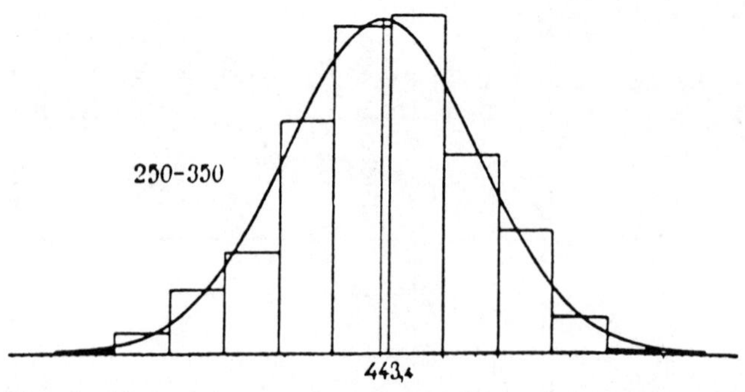

Fig. 6. The variation of seed-weight in the progeny of 1902 from the weight-class 250 - 350 mg., 1901 (835 beans).

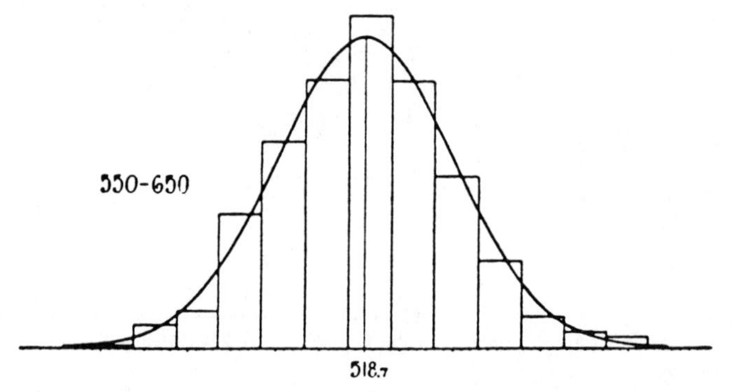

Fig. 7. The variation of the progeny 1902 from the weight-class 550 - 650 mg., 1901 (609 beans).

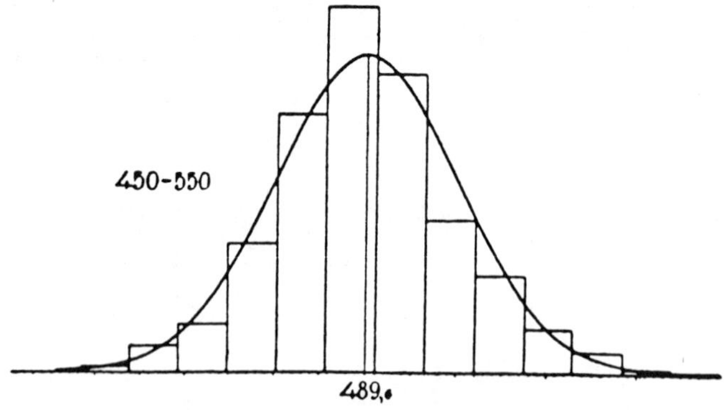

Fig. 8 The variation of the progeny 1902 from the weight-class 450 - 550 mg., 1901 (1138 beans).

SUMMARY TABLE 4

SEED WEIGHT OF BROWN BEANS, 1902. RELATIVE NUMBERS

The individual pure lines are considered separately, Within each line the average seed-weight is set = 100. The underlined figures state the number of individuals in question. See besides Summary Table 2. Concerning both columns to the right see the text.

Classification of the mother-seeds	150–200	200–250	250–300	300–350	350–400	400–450
Number of seeds sown	6	14	32	55	150	78
Line A	....	....	....	....	....	....
— B	....	....	....	....	103 86	....
— C	....	....	....	....	....	....
— D	....	....	....	....	....	....
— E	....	....	....	....	....	103 107
— F	....	....	....	101 117	....	....
— G	....	....	91 28	....	105 105	99 307
— H	....	....	....	99 114	103 81	98 136
— J	....	....	98 40	110 53	100 164	98 155
— K	....	105 18	....	....	....	....
— L	....	....	....	103 147	99 90	....
— M	....	103 78	....	....	99 217	....
— N	101 54	....	103 111	95 92	100 100	....
— O	....	....	102 72	....	99 147	....
— P	....	....	100 21	....	....	104 51
— Q	....	93 16	....	....	101 262	....
— R	....	109 14	....	....	....	....
— S	....	....	....	....	100 119	....
— T	....	....	106 20	....	100 111	....
Compilation of all lines with consideration of their numbers of individuals	99 54	103 126	101 292	101 543	101 1482	99 756
	101		101		100	

Generally, the agreement with the ideal-curve is fully as good as was the case with the pure lines. The greater number of individuals in the progeny series considered here compensates the irregularities;

450–500	500–550	550–600	600–650	650–700	700–750	Progeny of the Minus-variants	Progeny of the Plus-variants
70	52	42	31	38	6		
....	....	94 15	100 39	99 45	103 46	97	101
96 118	102 77	101 72	101 48	99 74	....	100	101
....	102 144	....	98 40	98 98	....	101	99
....	99 32	98 163	....	103 112	....	99	102
....	96 29	....	....	98 119	....	101	98
99 124	....	....	....	....	....	101	99
101 93	....	....	....	....	....	98	100
102 87	....	....	....	....	....	101	100
95 103	103 102	101 95	....	....	....	102	100
99 131	....	....	100 39	....	....	103	100
92 36	....	....	....	....	....	101	94
....	....	....	....	....	....	103	99
....	....	....	....	....	....	102	98
....	....	....	....	....	....	102	99
....	....	95 34	....	....	....	101	98
....	....	....	98 27	....	....	96	99
....	99 42	....	97 27	....	....	106	98
....	97 20	....	....	....	....	100	98
....	....	84 10	....	....	....	104	90
99 692	101 446	98 389	99 220	100 448	103 46	100.9	98.5
100		99		100			

nevertheless it is singular to see as beautiful Galton-curves as in Figs. 6 & 7 (p. 31) with a notoriously very heterogeneous material. If one did not know the real conditions, one would only too easily take these curves as a beautiful expression of a true displacement of type by means of selection, as for instance, Ammon (l. c.) assumes, and as must evidently be the opinion of Galton and Pearson that the effect of selection is generally to be understood. In our examples, however, we have to do only with exceedingly imperfect isolation of the types of the original population. It is just such an incomplete purification which explains that continued selection must bring about "improvement" in the direction of selection: this causes more and more complete purification, a sharper isolation of similarly-constituted lines. It is easily understood that an ordinary selection through a few generations can effect an almost (or occasionally completely) accomplished purification and that thereby, practically speaking, an end of the improvement in regard to the particular single character[1] will be reached. However, no guarantee at all will be given that those individuals which are only plus- (or minus-) variants of types other than that one which selection strives to "make" (i.e., isolate) will be excluded thereby. There lies in this a feature of importance in judging the often mentioned tendency of selection-races to "degenerate." This whole matter, which is by no means satisfactorily studied, I cannot discuss in more detail here.

We believe therefore, in complete agreement with all earlier experience, that selection must necessarily operate in populations—if that is, the population includes different types, and where is this not the case in nature?

But how does selection act within the individual pure lines? Can the type here be shifted? The tables of separate lines (pp. 17-21) give the information, that selection has had no reliably demonstrable influence on the types of the pure lines. This matter is most simply and clearly demonstrated by Summary Table 4 (pp. 32-33), which corresponds completely to Summary Table 2, except that in Table 4 all weight data are expressed not in absolute numbers, but in percentage of the average weight of the pure line in question. For the rest the figures can speak for themselves.

To the right in Table 4 are two columns, in which, for each line, the characteristics of the progeny (1902) of the plus- and of the minus-

1. Regarding this see de Vries, Mutationstheorie, Vol. 1, p. 62.

variants (1901) are grouped together. In the main table each individual seed has had equal influence in the calculations; in the grouping mentioned here each progeny-series (of a mother-class) was equivalent in the calculation. As one can see, this way of calculating has no essential effect on the result.

In several lines selection appears to have operated, in others the result was quite the opposite—on the whole, within the pure lines nothing at all has been achieved through selection. In the case of another variety of bean (a black Belgian bush-bean) I have obtained corresponding results; the data however were not large and should serve only as a check, therefore a closer consideration of the figures is here superfluous.

How, consequently, does the Galtonian regression law manifest itself within pure lines? The answer turns out to be clearly and plainly this: the regression is complete, quite up to the type of the line. The personal character of the mother-bean has no influence, that of the grandmother, etc., also none, but the type of the line determines the average character of the offspring—of course in joint effect with the whole life situation in the year in question.

[*Editor's Note:* Pages 40–56 of the original material were not translated.]

All that was reported here gives at the same time a full corroboration and a complete elucidation of the well-known regression law of Galton, concerning the relation between parents and offspring. Other regression relations are not under discussion here.

As far as my experimental data go, they agree very well with Galton's doctrine that individuals, deviating from the mean character of the population, beget offspring which—seen on the average—deviate in the same direction but in lesser degree. Therefore a selection in the population causes larger or smaller displacement—in the direction of the selection—of those mean characters around which the individuals in question fluctuatingly vary.

Since I did not stop at the point of examining the population as a unit but was able to resolve my material into its "pure lines," it has appeared in all cases that within the pure lines the regression has been, so to say, complete; selection within the pure lines has brought about no displacement of type.

The shifting of the mean characters which selection in populations can, it is well-known, generally produce, is accordingly caused by the fact that the given population—at least in my material—consists of different "lines," whose types can be more or less diverse. With the usual selection in populations the work is done uncleanly; the result depends upon incomplete isolation of those lines whose types deviate in the direction in question from the mean character of the population.

The usual well-known result of selection—successive progress in the direction of selection in the course of a few generations—depends accordingly on the progressive purification with each generation of the deviating line concerned. And it will now be easily understood that the action of selection cannot be carried out beyond fixed limits—it must indeed cease when the purification, the isolation of the particular most strongly deviating line, practically speaking, is carried to completion. In this connection it must be mentioned that one can never conclude with certainty, from the agreement of a table or curve of variation with the numerical relations of the exponential law of errors (or binomial formula), the existence of only one type in the variant material. The variation curve of the individuals of a one-strain population, in the usual sense, may frequently, perhaps in most cases, be the expression for the fact that numerous types are represented by the different lines of the population. The mean value then by no means signifies a true type. In this whole procedure the great defect of a purely statistical method appears.

Therefore, I have tried everywhere in this treatise to distinguish sharply between the concept mean (mean character, mean value, etc.) and the concept type.[1] The confusion of these two completely different concepts has only too often given rise to misunderstanding and false inference, and this not only in the theory of heredity. It must, however, be admitted that in individual cases it can often be very difficult, without closer analysis, to accomplish the sharp separation of the two concepts; and in pure lines the two ideas can very often have the same content. The numerical expression of a type is very frequently, though certainly not always, an average number.

In the case of morphological characters — at least with whole series of those characters whose value in classification will, for this very reason, be generally acknowledged — the distinction between different types is such that the single individual, in spite of all variability, can be acknowledged, mostly without further ado, as belonging to one or another of the narrowest systematic groups (e.g., the "little species" of Jordan).

The morphological types in question can as a rule scarcely be ordered to one variation series so smooth that a mixture of individuals of different types could be confused with a series of individuals belonging to one single type. A mixture of de Vries' Oenothera forms or of Raunkier's Taraxacum "races"[2] gives a different picture, in regard to the essential morphological characters, than does a pure culture of one single form.

In the case of all kinds of characters of a more physiological sort — the non-botanical characters of Hj. Nilsson — as, for example,

---

1. See note n. 1, p. 4.
2. C. Raunkiaer, Kimdannelse uden Befrugtning hos Maelkebotte. (Botan. Tidsskrift, Vol. XXV, Copenhagen, 1902, pp. 109-139.)

most size and other quantitative relations, chemical properties, certain numerical proportions, and so on, the picture is different. Here the different, actually existing types—easily demonstrable through isolated cultivation—show mostly only quantitative differences, so that the variation curves of different types can flow together; one has the transgressive curve of Hugo de Vries. A mixture of individuals which belong to well-separated types in regard to one of these characters (see small and large, or narrow and broad beans), can therefore very easily form a variation series so smooth that type differences are not to be detected directly, while the mean value can erroneously be regarded as a unique type. In these cases, it becomes impossible to decide from the single individual to which type it belongs. Table 3, p. 30, gives very good examples for illustration.

For all these reasons, more or less clearly recognized or sensed, the study of the first named, I would like to say "true," morphological characters has been the central point of systematics. Only recently did the more physiological characters become included in the sphere of interest of systematics, recognized especially in lower forms. It is in these relationships also, concerning the true morphological characters, that the mutation theory has had as yet its best supports in practice. On the other hand, these characters, which chiefly determine the whole habit of the plant, cannot, or can only partly, be expressed numerically, and their estimation falls almost always outside the range of the exact methods of measurement and calculation.

Biometrical research, i.e., the exact investigation of the laws of variation and of heredity, has therefore kept first of all to the more physiological characters or on the whole to the characters called "meristic" by Bateson, those which lend themselves to expression clearly in number, measurement, or weight. And here, where diversity of type in the comparison between individual and individual is usually not at all to be differentiated from the manifestations of fluctuating variability, the Galton-Pearsonian conception has, naturally, its firmest stronghold; here one must—by disregarding the pure lines—necessarily come to the result that a selection of deviating individuals (plus- or minus-variants) can cause a real shifting of the type at hand. It is apparently this till now wholly legitimately existing conception which must have hindered the biometricians, represented by Weldon and Pearson, from accepting mutations as something other and more important than fluctuating variations. And with the hitherto existing statistical findings concerning inheritance in populations the acceptance of mutation was perhaps not essential for biology. I say here "perhaps" in order to come part way to meet the biometricians. For myself. the brilliant researches of Hugo de Vries have long removed every doubt concerning the existence of mutations.

After my results given here, it will indeed appear that the basis of the Galton-Pearsonian laws, the relation between parents and offspring, is a somewhat different one than we were previously disposed to accept. The personal nature of the parents, grandparents, or any ancestor whatsoever has, so far as my experience extends, no influence on the mean character of the offspring. But it is the type of the line which determines the mean character of the individuals, of course in intimate co-operation with the influence of the external environment of the specific place and time. The "line" is in so far "completely constant and highly variable," as de Vries, so characteristically, and only seemingly paradoxically, expressed a similar behavior.[1]

With this, however, let it by no means be said that one supposes the pure lines to be absolutely constant.

First, it is a possibility that a selection of fluctuating variants through very many generations can yet ultimately shift the type of a line. Nothing positive supports this — the statements of the biometrician concern, as often said, populations which are not split up, or rather cannot be split up, into pure lines. The burden of proof will here rest on him who wants to assert a selection of that kind.

Secondly, we have to think about hybridizations — but thereby the line ceases to be pure! The whole hybrid question is not relevant here.

Thirdly, however, there are mutations, the possibility of erratic changes of the type. To explain them would be most premature; their existence must first be established to much wider extent than has so far been done. That they occur is to me, as said, certain; I hope in a later work to be able to report my own positive findings.

1. Die Mutationstheorie, Vol. 1, p. 97.

Let it be said here, that a mutation in a given direction cannot be specifically awaited in the descendants of individuals who deviate fluctuatingly in this direction.

At this point I must touch on the delicate question, which can illustrate the statement of de Vries, that one frequently first observes minus-variants of the newly appearing type—a matter which, not without reason, has roused the scepticism of the biometrician. It is to be hoped that continued studies will bring light to this matter which only apparently effaces the boundaries between fluctuating variability and mutation.

(Addendum: In the closing section of his Mutationstheorie—which was only published after the editing of this work—de Vries (l.c., p. 503-504) has reasoned ingeniously why in most cases mutations first crop up in hybrid forms. Therein lies an instance exceedingly important to the explanation of the relations just mentioned.)

Hugo de Vries, in his Mutationstheorie (Vol. 1, p. 368 ff.), has a special chapter on "Nutrition and Selection," in which particularly the after-effect of the plentiful or scanty nutrition of a mother plant is discussed. I do not doubt that different findings, be they real or supposed, which have been made in regard to the effect or the lack of effect of a selection, can find their explanation by means of the reason preferred by de Vries. And of special interest is the phase of ontogenesis called "sensitive period" by de Vries. In my material presented here I have nevertheless scarcely a point of contact with that just cited. Of course, it is absolutely not my intention to wish to explain directly through the "principle of pure lines" all forms of the shifting of character which a selection, in conjunction with extreme or specially produced living conditions, might produce. In this direction, which strongly interests the neo-Lamarckians, there is still very much to explain—and especially with the use of truly pure lines as starting material.

My task was here, first of all, only to clarify the Galtonian regression between parents and children, and here I believe that my material, which obviously has a natural character just as Galton's, has its value as basis of an analysis of the Galtonian laws valid for populations. My data as such conflict neither with Galton's data nor with de Vries' presentation.

If my investigations are accurate, and if their significance spreads beyond the particular special cases, the general results of this work will form a not unimportant support for the doctrine, at present especially represented by Bateson and de Vries, of the great significance of "discontinuous" variation or "mutation" for the theory of heredity. For a selection in populations acts in my cases only in so far as it chooses representatives of already existing types. These types are not successively formed, perhaps through the protection of those individuals which fluctuatingly vary in the specific direction; but they are found and isolated.

In the study of heredity in populations in which, owing to regular necessary cross-fertilization or hybridization, pure lines cannot be isolated at all, the findings gained by the study of pure lines must nevertheless serve as ground work—here however combined with knowledge of hybrid theory. This conception stands, more closely regarded, in full agreement with the basic ideas in the great, often mentioned works of de Vries—as one sees, the conception for my part was reached by a somewhat different way than that followed by de Vries, further also, and most significantly, supported by somewhat different facts.

Also the important question concerning correlative variability receives a somewhat different character according to whether one works with pure lines or with populations. In the latter case, a given degree of correlation (Pearson's "ratio of correlation") will not necessarily represent a really strong regularity, as I have tried to clarify earlier. Within the pure lines the validity now of the discovered ratio of correlation is, however, all the greater. Summary Table 6 (p. 43) speaks quite definitely for this concept, since it was not at all possible to alter, through selection within the pure lines, the correlation between length and breadth of the beans, while it was easy to isolate different types, e.g., narrow and wide forms, from the existing, at first seemingly quite homogeneous, populations.

But we have also to reckon here with the possibility of mutations, through which the strongest correlation ratio can be broken. However this question is not yet under consideration here; in a later work I hope to be able to examine it more closely and moreover with the principle of pure lines as basis of the research.

I would very much regret should the reader have received the impression that the value of the important works of Galton, Pearson, and the rest of the biometrical investigators is questioned here. The treatment which Pearson, in particular, has given the question of the ancestral influences within a given population I am not emboldened to criticize. I believe, however, that the principle of pure lines in the hand of a Pearson can lead the biometrical study much further than can his study of populations. Of course the ratios studied by Pearson have their great scientific importance and they have great practical interest at the same time—but they are not adapted to throw full light upon the fundamental laws of inheritance.

And as for the investigations of Galton, I do not see otherwise than that the results and concepts submitted here support in the most beautiful manner the basic ideas of the Galtonian "stirp"-theory,[1] already carried out in 1876.* This theory held almost all that had real value in the later theory of Weismann of the "continuity of the germ-plasm." The fact that Weismann's speculations[2] could over-shadow the more simply presented but thereby not less ingenious and everywhere original ideas of Galton is due in part to the fact that Galton himself in his later publications has neither determinedly adhered to the stirp-theory nor adjusted it to progressing research. The stirp-theory does not agree well with the Galtonian regression law—but it can hardly be better supported and illustrated than through results such as those reported here: an, <u>on the average, complete regression</u>

1. Galton's theory is well known to me from his original paper in <u>Revue scientifique</u>, Vol. X, 1876, p. 198, (Theorie de l'hérédité.)

* [In Pearson's words, "proceeding from the idea of the reproductive cells only being the bearers of hereditary characters, Galton evolved the conception that what we now term the germ-plasm of the individual is not peculiar to the individual but to his 'stirp.' Individuals are merely the conduits through which flow the germ-plasm of the stock or stirp to the next generation. The child is not like the parent because it springs from the parent, but because both are to a definite extent representatives of the same stirp, i.e., partial or in the case of parthenogenesis complete products of the same germ-plasm." Trans.]

2. The attitude of the biometricians to "Weismannism" is clearly presented through Pearson's characterization of this trend in "Socialism and Natural Selection" (<u>Fortnightly Review</u>, July 1894). Printed in Pearson's <u>Chances of Death</u>, Vol. 1, 1897, p. 104. Here, however, I do not have to deal with this question more closely.

to the type of the line seems to me the finest evidence for the justification of the Galtonian stirp-theory. To be sure Galton's stirp-theory cannot be maintained unchanged. While Weismann, just as Galton, still in recent times[1] lets the organs or rather cell regions be represented through "determinants"—or whatever one wants to call these theoretical hereditary corpuscles—de Vries has the great merit of having established the properties (single properties) as units, represented through "pangenes"—a conception which dates from 1889[2] and is further reasoned in the *Mutationstheorie.* It seems to me that the Galton - de Vries theory is for the moment the only useful—and moreover very useful—theory of heredity.

Should it be possible, through the work submitted here, to bring the principle of pure lines to recognition as an absolutely necessary principle in really thorough-going investigation in the sphere of hereditary theory, the main goal of the publication would be reached. Later publications will then examine more closely the behavior of lines which vary polymodally, whereas here only unimodally varying lines were considered[3]—in order to present first of all my conception by means of the simplest cases.

The train of thought which was the basis of these investigations is most clearly expressed in its simplicity with the often quoted words of Goethe:

> "Dich im Unendlichen zu finden
> Musst unterscheiden und dann verbinden."

Vilmorin has emphasized the differentiating; Galton taught us the combining according to laws; what I have attempted here is: to combine the viewpoints for which these two gifted investigators have the reputation.

At the close of this work, which I have placed before the Königlichen Dänischen Gesellschaft der Wissenschaften in the meeting of 6 February 1903,[4] I must give my loyal co-workers a warm thanks,

1. Weismann, *Vorträge über der Descendenztheorie*, 1902, p. 421.
2. De Vries, *Intracellulare Pangenesis*, Jena 1889.
3. De Vries in the *Mutationstheorie* (Vol. II, p. 509) has given a special case of heredity in a pure "bimodal" line, based on my reports by letter.
4. The paper is published in Danish in *Oversigt over det k.d. Videnskabernes Selskabs Forhandlinger 1903*, No. 3, (July 1903).

particularly Herr Dr. phil. Kölpin-Ravn, Mag. scient. A. Didricksen, and cand. hort. H. Stenboeck. And without special financial support on the part of the Carlsbergfond and the Königlichen Dänischen Landwirtschaftlichen Hochschule, it would have been impossible for me to carry out the extensive investigations whose first result is here presented.

19

Reprinted from *Biometrika* **2**:499–503 (1903)

INHERITANCE IN *PHASEOLUS VULGARIS*

W. Weldon and K. Pearson

Professor F. Johannsen has just published a summary of his recent experiments on inheritance. His work *Ueber Erblichkeit in Populationen und in reinen Linien* (Fischer, Jena, 1903) is dedicated to Francis Galton, as the "Schöpfer der exakten Erblichkeitslehre"; it shows that the author has realised the importance of adequate statistical methods in any attempt to deal with the problem of inheritance, and we wish to express our gratitude to him for the courteous tone adopted in speaking of "Biometriker," and for the patient effort he has made to understand their work.

Professor Johannsen's material is collected from three generations of beans: he has (1) a series of individual seeds, chosen out of a sample of about 16,000, which had been harvested and mixed together; these afford his evidence concerning the character of a grandmaternal generation; (2) each grandmaternal bean, when sown, yielded a crop of maternal beans, the mean character (weight or length-breadth index) of those borne by one plant being taken as the measure of the maternal character of the line of ancestry to which the plant belongs; (3) from the series of seeds borne by each mother plant, a certain number (from two to seven) were sown, the mean characters of the seeds borne by the resulting plants being taken as a measure of the characters studied in the filial generation of each line.

From these data three principal sets of tables are constructed; on p. 25, *Uebersichtstabelle* I. gives material for measuring the regression of individual filial beans on the mean weight of the maternal beans; a series of tables on pp. 21—24 gives the relation between the weights of the individual beans of the maternal plants, which were sown, and the mean weight of the seed produced by every resultant plant; finally on pp. 36—37 a table is given which Professor Johannsen believes to show that the progeny of every seed, within a particular line of ancestry, exhibits a complete "Rückschlag" to the type of its line, the coefficients of correlation and regression between parent and offspring within the line being each = 0.

On the basis of these tables Professor Johannsen attempts to explain the apparent discrepancy between Galton's law of regression and the results obtained by de Vries and others: but his view of the consequences supposed by "Biometriker" to follow from Galton's results shows that he has not fully realised what those consequences are.

It is fully realised by "Biometriker" that the general regression observed when we compare a filial generation with a parental generation is compounded of a series of sub-regressions, the members of each line of ancestry regressing to the "type" of their line; the effect of selecting definite ancestry for a small number of generations is also recognised; these points were fully dealt with in 1898, although few biologists seem to have realised the fact; it was then said:

"We now see that with the law of ancestral heredity......a race with six generations of "selection will breed within 1·2 per cent. of truth ever afterwards,"* or in other words the

* K. Pearson: "On the Law of Ancestral Heredity," *Roy. Soc. Proc.* Vol. 62, 1898. Cf. pp. 397—402, "On the Variability and Stability of Selected Stock."

fixity of a line of ancestry is asserted when the ancestral purity is far less than that involved in Professor Johannsen's "reine Linie," which he defines as consisting of "Individuen, welche "von einem einzelnen selbst befruchtenden Individuum abstammen" (p. 9).

Again, "with a view to reducing the *absolute* variability of a species it is idle to select beyond "grandparents, and hardly profitable to select beyond parents. The ratio of the variability of "pedigree stock to the general population decreases 10 per cent. on the selection of parents, and "only 11 per cent. on the additional selection of grandparents. Beyond this no sensible change "is made."*

In these two passages the fixity of the type and the high variability of individuals about the type are asserted as absolutely as they are asserted by de Vries in the passage of the *Mutationstheorie* quoted by Professor Johannsen. With our present knowledge of the coefficients of inheritance in man, horse, and dog, it would seem that from 2 to 4 generations of selection suffice to form a line varying greatly about its type, yet remaining true to that type. When we are told that a bean breeds true to its line we are told something which has been shown to be a necessary consequence of the law of ancestral heredity; if it were not true, the whole law would be upset.

The difference between the view put forward by Professor Johannsen, and that expressed in 1898 in the paper "On the Law of Ancestral Inheritance" is therefore not a difference which concerns the focus of regression in the offspring of selected ancestors; it is simply a difference as to the relation between successive generations of individuals within the line of ancestry. Professor Johannsen believes the tables on pp. 21—24 of his work to show that "*die persönliche* "*Beschaffenheit der Eltern, Grosseltern, oder irgend eines Ahnen hat*—soweit meine Erfahrung "reicht—*keinen Einfluss auf den durchschnittlichen Charakter der Nachkommen.* Es ist aber "*der Typus der Linie*, welche den durchschnittlichen Character der Individuen bestimmt......" (pp. 61—62). Within the same line of ancestry, whatever individuals of a generation be chosen as parents, the character of the resultant filial generation will be the same according to Professor Johannsen, or the coefficients of correlation and regression between parents and offspring, within the same line of ancestry, will each = 0.

The experimental results do not seem to us consistent with this view. If the offspring of every generation, within a given line of ancestry, breed true to the type of their line, subject to such seasonal and climatic influences as affect the whole generation of their year, then when the whole filial generation is compared with the whole parental generation, correlation between the two must necessarily be perfect, and the coefficient of regression must be simply the ratio between the standard deviations of the two generations. Professor Johannsen should, we think, first have shown that perfect correlation does in fact exist between parents and offspring in two successive generations of plants; and this he has failed to do; in the case of his maternal and filial generations he has, however, published data which enable us to determine the required correlation, and the table below gives the result. With such data, the only method available is the first method used for Shirley Poppies†. It consists in determining the correlation between *every individual bean* of the filial generation, and the *mean character of its parent.* The absolute value of the correlation so obtained will not be significant, but the coefficient of regression will closely represent the true parental correlation,—being a relation between mean filial character and mean parental character. Taking Professor Johannsen's *Uebersichtstabelle* I. (p. 25), and the maternal means given on pp. 21—24, the following table has been constructed, giving a coefficient of correlation = 0·531 and of regression = 0·591 ± 0·125. This latter value represents the coefficient of correlation, so far as the data allow it to be determined, between filial and maternal plants; and considering the paucity of maternal plants (only 19) the result is not in bad accord with previous results for parental correlation‡. The value 0·591 ± 0·125 is not very divergent from 0·5, but it cannot be held to approximate to unity!

* Pearson: *loc. cit.*

† *Biometrika*, Vol. II. Part I. p. 69.

‡ *Biometrika*, Vol. II. p. 379.

TABLE A.

Correlation between Individual Filial Beans and Mean Maternal Beans.

Filial Beans.

Mean Maternal Beans — Weights in milligrams	*10—15*	*15—20*	*20—25*	*25—30*	*30—35*	*35—40*	*40—45*	*45—50*	*50—55*	*55—60*	*60—65*	*65—70*	*70—75*	*75—80*	*80—85*	*85—90*	Totals
A & *D 600*	—	—	—	5	2	11	26	47	82	98	84	46	26	17	6	2	452
C 570	—	—	—	—	—	5	14	50	76	58	44	29	5	1	—	—	282
B 520	—	—	—	1	6	19	32	66	88	100	90	50	19	1	3	—	475
E 512	—	—	—	4	1	12	29	62	65	57	19	6	—	—	—	—	255
K 510	—	—	1	2	6	31	55	55	28	6	4	—	—	—	—	—	188
Q 440	—	—	1	2	7	16	44	93	80	52	10	—	—	—	—	—	305
R 410	—	—	—	2	3	12	17	27	19	3	—	—	—	—	—	—	83
S 405	—	—	1	2	3	8	27	47	37	30	4	—	—	—	—	—	159
G & *J 400*	—	1	5	23	66	155	283	353	241	97	15	—	1	5	—	—	1245
F & *T 395*	—	—	—	2	9	27	66	111	85	58	22	1	1	—	—	—	382
P 390	—	—	—	3	1	18	35	27	13	3	4	2	—	—	—	—	106
H 380	—	—	1	6	20	60	106	114	75	33	3	—	—	—	—	—	418
L 360	—	—	1	5	15	37	88	76	33	13	4	1	—	—	—	—	273
M 340	—	—	4	9	26	56	82	76	32	9	1	—	—	—	—	—	295
N 312	1	3	11	22	29	72	120	69	23	5	2	—	—	—	—	—	357
O 310	4	4	5	19	69	69	44	5	—	—	—	—	—	—	—	—	219
Totals	5	8	30	107	263	608	1068	1278	977	622	306	135	52	24	9	2	5494

The letters *A* to *T* are used by Professor Johannsen to denote his nineteen "pure lines."

If Professor Johannsen's hypothesis were true, the only way in which he could account for the regression here observed,—a regression whose existence he himself admits,—would be by assuming that the characters, by which he has described the plants of his pure lines, are imperfectly correlated with the actual mean characters of his plants which he supposes to represent the types of his lines; but this assumption, while leaving his hypothesis logically unshaken, would destroy the whole value of his experiments as evidence in its favour, by destroying the value of the measure by which he has determined the characters studied.

Some of the difficulties we have felt in following Professor Johannsen are undoubtedly due to the imperfect way in which he has measured the characters of ancestral plants*. The *grand-maternal plants*, for example, are determined each by the character of a single bean; the small value of such a determination may be judged from the variability among the beans of a single plant produced by the last generation; the mean number of such beans was 84·5, and the mean standard deviation of all arrays due each to a single plant was 75·37 milligrams, so that the mean character of a mother-plant, inferred from a single bean chosen at random among the offspring, would be equally likely to lie inside or outside the limits (true maternal mean + 50 mgrm.) and (true maternal mean − 50 mgrm.). There is thus, we venture to say, no strong probability that the numbers by which Professor Johannsen describes his grandparental beans represent the mean character of the seeds of the corresponding plants within ± 100 mgrm.

Again, the whole evidence, that the coefficient of filial regression within the line is zero, rests on the tables on pp. 21—24; but in these tables we are only told (1) the mean weight of seeds

* A preliminary study of homotyposis in the bean must of necessity precede any attempt to measure plant character by a single seed. Professor Johannsen would have to show that the homotypic correlation was perfect to justify his measure. This is very far from the fact not only in *Phaseolus vulgaris*, but in all beans hitherto examined from this standpoint.

produced by a mother-plant, representing the maternal character in a line; (2) the weights of individual seeds, taken from this mother-plant and sown; and (3) the mean weight of the seeds borne by each resultant plant. The regression coefficient, which Professor Johannsen regards (without any very adequate proof) as equal to 0, is between the deviation of the *individual beans* (2) and that of the mean character of the series of plants (3) resulting from them. We have therefore (1) the maternal generation defined by the mean character of all the beans borne by a plant; (2) a second generation, children of the maternal plant, each defined in the same way, by the mean character of the beans it bears; we have no third generation at all, and the regression which Professor Johannsen has observed seems to have little bearing on the question at issue, which could only be determined by growing a third generation from representative offspring of the filial plants, describing each plant of this generation as those of the two previous generations were described, in terms of the mean character of its seeds, and then determining the correlation between the characters so described in the two successive generations, the children and grand-children of the single plant originally used to determine the line.

The question, which the tables given do to some extent answer, is the question what relation exists between the character of two seeds from the same mother-plant, and the character of the plants produced when those seeds are sown. Now it seems clear that if we take small beans out of a general harvest of seed, we shall be to some extent selecting the seeds of plants which bore on an average small beans; but what reason is there for supposing that the small and the large seeds from one and the same plant will lead to groups of plants bearing respectively small and large seeds? The hypothesis involved in this supposition seems somewhat analogous to the view that out of two eggs of a clutch, the smaller will produce a hen laying smaller eggs than that produced from the larger; this may well be quite fallacious, and it may yet be true that out of large masses of eggs, small eggs produce on the whole hens which lay small eggs. The absence of relationship of marked kind between the weight of seed sown, and mean weight of seed produced by the resulting plant, seems to have no bearing on the problem whether selection within the line can produce a change of character.

One further point we must notice; the *Uebersichtstabelle* 4 (pp. 36—37) has been treated in a quite illegitimate way, which would make the coefficients of correlation and regression = 0 between any two variables whatever.

If, as Professor Johannsen believes, the individual differences between members of any one generation were mere fluctuations, having no hereditary value, then a given generation ought as he says to be as well determined by selection of its grandparents as by selection of its parents. We cannot determine whether this is true of the average character of plants in Professor Johannsen's experiments, because the necessary data are wanting; but we can determine roughly the relation between three successive generations of *individual beans*. The material has been selected in such a way that the standard deviations of the successive generations have clearly quite artificial values, so that the correlations obtained are not very trustworthy; further, the exact weights of beans are not always given, so that we have been obliged to place a bean recorded as lying between say 400 and 450 mgrm. in the middle of its category. With these qualifications, we find

Correlation of Mother Bean and Offspring Bean $r_{01} = 0{\cdot}3481 \pm 0{\cdot}0080$,

Correlation of Grandmaternal Bean and Offspring Bean $r_{02} = 0{\cdot}2428 \pm 0{\cdot}0086$.

If we wish to predict the weight of a given bean of the filial generation from the known weight of its maternal bean, we must form a regression equation, which becomes

Probable weight of Offspring bean $= 538{\cdot}31 + 0{\cdot}2691 \times$ weight of Maternal Bean ....(i).

If we wish to predict the weight of a filial bean from knowledge of the grandmaternal bean, we obtain the regression equation

Probable weight of Offspring bean $= 417{\cdot}41 + 0{\cdot}1074 \times$ weight of Grandmaternal Bean...(ii).

This shows us at once that from Professor Johannsen's own data the maternal bean is more than twice as influential as the grandmaternal bean in settling the weight of the filial generation.

If we correlate the selected maternal and grandmaternal beans, we find the correlation, $r_{12} = 0{\cdot}2532$; and from this and the preceding correlations we find the double regression formula

Probable weight of Offspring bean

$= 330{\cdot}71 + 0{\cdot}2373 \times$ weight of Maternal Bean $+ 0{\cdot}0731 \times$ weight of Grandmaternal Bean...(iii),

showing again the predominant influence of the maternal bean.

If we were to calculate the mean weights of each array of offspring means from (i), (ii) and (iii), we should expect the mean errors of the results to be in the ratio of

$$\sqrt{1 - r^2_{01}},\ \sqrt{1 - r^2_{02}} \text{ and } \sqrt{\frac{1 - r^2_{01} - r^2_{02} - r^2_{12} + 2r_{01}r_{02}r_{12}}{1 - r^2_{12}}},$$

or in this case as

$$1{\cdot}014 : 1{\cdot}050 : 1.$$

We have applied (i) (ii) and (iii) to the 65 arrays of offspring given by Professor Johannsen, and the mean errors are

$$44{\cdot}3;\ 45{\cdot}5, \text{ and } 42{\cdot}8$$

or in the ratio of

$$1{\cdot}035 : 1{\cdot}060 : 1.$$

These numbers are, perhaps, as close as we could expect, and they show that we do in fact get better results from a knowledge of maternal bean than from knowledge of grandmaternal bean, and better results from a knowledge of both together than from a knowledge of either alone.

We hold therefore that Professor Johannsen's results prove:

(1) That there is a regression from parent to offspring, leading to the inference that parental correlation has for *Phaseolus vulgaris* a value closely identical with that found for other animals and plants, when we compare mean parental and mean filial characters;

(2) That when we compare the characters of individual seeds in successive generations the correlation between a seed and its parental seed is so much greater than that between a seed and its grandparental seed (both belonging to the same pure ancestral line) as to give strong evidence that characters arising in one generation within the line are inherited, and do therefore afford a basis on which selection may act.

W. F. R. W. AND K. P.

20

Reprinted from *New Phytol.* **2**:235–242 (1903)

PROFESSOR JOHANNSEN'S EXPERIMENTS IN HEREDITY: A REVIEW.

BY G. UDNY YULE.

THE statistical theory of heredity, as developed in the work of Galton and Pearson, concerns itself with aggregates or groups of the population and not with single individuals. A number of laws have been formulated, on the basis of researches conducted in accordance with such theory, of which the most important is Galton's Law of Regression, *i.e.* the law that the offspring of abnormal parents are *on the average* less abnormal than such parents, "regressing" towards the mean of the race.

But the student of heredity, says Professor Johannsen[1] (p. 4), "cannot rest satisfied with such essentially statistical researches. A race, a population, an aggregate (Bestand) of any sort . . . . is by no means always, from the point of view of the biologist, to be treated as homogeneous (als Einheit), even when the individual variations are grouped round an average value, presumably "typical," in the manner expressed by the law of error. Such a population may . . . . . . contain a number of independent types (selbständige Typen), differing markedly from one another, which may not be discoverable at all by direct observation of the empirical frequency curve or table."

"Before such a population is treated as homogeneous" he continues (p. 5) "it should therefore be biologically analysed in order to be clear as to its elements, that is to have some knowledge as to the independent types (selbständige Typen) already existing in the population. Only after such an analysis is one in a position to decide whether, and how far, a treatment of the material as homogeneous is permissible. It is in fact such an analysis that Louis Leveque de Vilmorin had in mind when he . . . . . laid down his 'principle of isolation' . . . . . According to this principle one must decide by isolation—that is by keeping separate the seeds of each individual mother plant—whether the 'force héréditaire,' as Vilmorin terms it, be large or small."

[1] Ueber Erblichkeit in Populationen und in reinen Linien, ein Beitrag zur Beleuchtung schwebender Selektionsfragen: von W. Johannsen, Professor der Pflanzenphysiologie an der kgl. dänischen landw. Hochschule in Kopenhagen. Fischer, Jena 1903 (pp. 68).

These passages give the aim of Professor Johannsen's work, *viz.* the elucidation of the statistical laws of heredity for the race by the study of the corresponding laws for the "pure line," *i.e.* the posterity of a single self-fertilised individual. The results of three researches are given, the first two referring respectively to the weight and the length-breadth ratio of the seeds of *Phaseolus vulgaris*, the third to the phenomenon of relative sterility in barley as measured by the percentage of buds in a head failing to set seed. The barley used was a variety of three-rowed chevalier-barley.

For details the reader must of course refer to the original, but the method of the chief experiments may be illustrated by reference to those on the weight of beans. A single bean of known weight was taken, sown, the resulting plant allowed to self-fertilize (in a net-covered enclosure) and its seeds harvested. These seeds or a sample of them were sown in their turn, allowed to self-fertilize under similar conditions, and their seeds harvested. The mother seeds and their offspring were then weighed, and the results collected in the following form. (All weights are stated in milligrams).

LINE A.

Weight of the ancestral bean ca. 800.
Average weight of the mother-beans ca. 600.

Weight of the mother-beans.	Characters of the Offspring.		
	Average weight.	Number.	Standard deviation.
550—600	605	15	126·8
600—650	642	39	107·9
650—700	635	45	105·8
700—750	661	46	112·4
The whole line	641·9	145	109·5

Details are given for nineteen such "pure lines" as regards the weight of the beans and for twelve lines as regards the length-breadth ratio; the barley is treated in much less detail. The results are very simply summarised by dividing each set of "mother-beans" into two classes, "minus-variants" and "plus-variants," and expressing the mean weight of the offspring of each of these classes as a percentage of the mean of the line. The results for all the different lines can then be grouped together (Table 4, p. 37).

As regards the weight of the beans Professor Johannsen finds

Off-spring of the	
Minus variants	Plus variants
100.9	98·5

this result being based on 5494 seeds ; and as regards the length breadth ratio (2440 seeds)

Off-spring of the	
Minus variants	Plus variants
99·8	100 0

In short (p. 39) "the regression is complete, right back to the type of the line. The individual character of the mother-bean has no influence, nor has that of the grandmother, but the type of the line (der Typus der Linie)—of course in conjunction with the whole environment in the year concerned—determines the average character of the offspring." While "in some lines selection appears to be effective, in others the result is the reverse of what one would expect—on the whole nothing whatever is attained by the process of selection within the pure lines." The writer does not, of course, dispute the effective character of selection in an ordinary population, but in such a case (summary p. 58), "the work is carried out on impure material : the result depends on the more or less complete isolation of those lines whose types deviate in the required direction from the mean of the population.......And it will consequently be easily understood that the operation of selection cannot be carried beyond a certain limit—it must cease when the purification, the isolation, of the lines which deviate most strongly in the required direction is, practically speaking, complete." Later the author adds, to further emphasise the point (p. 64), "selection in a population is only effective in many cases in so far as it picks out the representatives of already existing types. These types are not formed in succession, *e.g.* by protection of those individuals that exhibit fluctuating variations in the required direction; they are simply found and isolated." These extracts give, I think, a fair idea of Professor Johannsen's position; but quotations removed from their context are never quite satisfactory, and I hope the reader will refer to the original.

The researches have been carried out on novel and carefully considered lines, and the results obtained are clearly of the highest importance for both practice and theory. As regards the former, they show conclusively that the breeder should proceed by noting the separate lines and not merely the offspring *en masse* of the individuals showing desired characteristics; by such a procedure he will attain any desired end with greater certainty and speed. As regards the latter, the fact that heredity may be vanishingly small within the pure line, although quite sensible within the population at large is a very striking fact and, I believe, a new one. It must find its proper place and explanation in any complete theory of intraracial heredity.

Professor Johannsen's explanation, as I understand it, is this, that the race consists of a number of distinct "types," in each of which the germ-plasm structure (or whatever we choose to term the character of the germ cell that determines the character of the soma) is unalterable; that in a pure line all the variations are consequently *purely somatic* and therefore non-heritable; and that selection cannot operate in such a race because there are no germinal variations to afford the necessary material. On this hypothesis the heredity within the pure race is absolutely zero.

The hypothesis seems, however, to be somewhat qualified by later passages. The meaning of the word "Typus"[1] is not defined. I have assumed that the writer means an *unalterable* germinal structure, as it is only on such an assumption that the conclusions seem to follow. But on pp 61-62 of the summary, after a repetition of the statement to the effect that it is the "type of the line" in conjunction with the environment which determines the character of the offspring, and an approving quotation of De Vries's dictum that the line is "völlig konstant und höchst variabel" (germinally constant, I take it, and somatically variable) we read "Damit sei aber durchaus nicht gesagt, dass die reinen Linien absolut constant sein sollen."[2] In the first place, it is suggested,

[1] Statistically, the word is used in two quite distinct senses, neither carrying any implication of stability in the biological sense. (1) One may speak of the type of the race as regards any given character, meaning thereby the modal or most frequent value of the character (identical with the average, if, but only if, the frequency distribution be symmetrical). (2) One may speak of organisms "of a given type" within the race, meaning thereby simply organisms possessing a given value of the assigned character, within more or less narrow limits.

[2] I met with precisely the same difficulty in reading, some months ago, the very interesting work of Klebs on "Willkürliche Entwickelungsänderungen bei Pflanzen." On p. 145 is stated "Da wir von der Voraussetzung ausgehen, dass jeder Spezies eine konstante spezifische Struktur zu Grunde liegt . . . . ." while three pages later we find "Die Aufgabe der Speziesphysiologie ist aber noch aus einem zweiten Grunde grenzenlos weil die Konstanz der spezifischen Struktur . . . . . . doch nur relativ ist." The fact however that it is "relative" (*i.e.* really not constant at all) would have been better stated earlier. It limits many of the author's conclusions to an extent he does not realise.

continued selection for *many* generations may lead to an alteration of type, although selection for one generation only gives no sensible effect. "The burden of proof lies however with those who assume such an effect of selection." In the second place one must consider "Mutations, or the possibility of discontinuous variations of the type." "That they do occur" says Professor Johannsen, "appears to me beyond doubt." The latter statement seems to indicate that the writer does not consider the existence of mutations would invalidate his theory.

But, surely, the truth or otherwise of the hypothesis, that continuous selection within the pure race will ultimately affect the type, is quite independent of the *nature* of variation? If variations—germinal variations—in the required direction arise *in any way*, that is all that is needful. If *mutations* occur they will be picked out by the selector in the first instance, and the proportion of mutants to somatic variants will increase as the selection is repeated, because all (or a large proportion) of the offspring of the former will be retained and only a small (or smaller) proportion of the progeny of the latter. Is not the denial of the possibility of effecting a change of type, by selection within the pure race, a denial of the possibility of evolution itself? The existence or non-existence of the effect is not merely a criterion—nor a criterion at all—as to the nature of variation.

It seems difficult then, on very wide grounds, to admit that the effect of selection within a "pure line" or the intensity of heredity within such a line can be rigidly zero, *i.e.* the "burden of proof" lies with those who hold such a conception, which is inconsistent with the conception of evolution itself. All that can be proved by such experiments as Professor Johannsen's is that the effect is small (compared with the probable error of the result)—an interesting result, but a very different matter; for if the effect in given cases be not *zero* but only *small* it may in other cases be *sensible*. If Professor Johannsen believes in the occurrence of mutations he ought to believe in the effect of continued selection within the race, whether accepting the hypothesis of continuous variation or not.

It is unkind to "ask for more" where so much is already given, but one may point out that it would be of the highest value for comparative purposes to have data of similar form for a character more strongly inherited. Would it be possible, for instance, without making additional experiments, to regroup the material so as to deal not with the weight of the *single parental seed*, but with the *average weight of the single seed on the parental plant* and the same character for the offspring? One would expect such a mode of grouping to give a greater intensity of heredity in the race at large and probably a sensible inheritance within the pure race.

The main question therefore seems to be this, are the results of Professor Johannsen's experiments consistent with the conception of continuous variation in the character of the germ-plasm? (That they are consistent with some form of discontinuous variation may, I think, be at once conceded). This question must be answered in the affirmative.

On the conception of continuous plasmic variation one may picture the formation of the germ cells of a pure race as accompanied by a process of gradual breaking up of the original germinal characters, controlled only by the action of selection (in the widest sense of that term). There must be a breaking-up of the original germinal characters (such as might be caused by the daughter cells receiving varying samples of the original germinal material), or no heritable variations would ever arise. There must be selection, or the breaking-up process would in course of time create variations deviating to an unlimited extent in every direction; which is absurd. Under the action of selection the process results in the genesis, from a single cell, of a series of somatic generations in which the variation gradually increases but asymptotes (provided all conditions are kept constant) towards a fixed limit. Under such circumstances the coefficient of correlation will also gradually increase and asymptote towards a fixed value, but the initial value and the ultimate value depend (1) on the intensity of selection (2) on the intensity of effect of circumstance—definite or indefinite—in producing *divergent* somatic characters from *similar* germ cells. The greater the selection, and the greater the divergence, the lower both the initial and ultimate values of the coefficient of correlation measuring the intensity of heredity.

To determine these two factors two independent data of some sort are needed, *e g.* the coefficients of correlation between mother and offspring, and between grandmother and offspring, in the ultimate race. Unfortunately no suitable data appear to be given by Professor Johannsen for the characters with which he deals. From two tables he gives (p. 16 and p. 42) one may however roughly estimate the coefficients of regression (for offspring on parents) for weight and length-breadth ratio of beans as about ·17 and ·23 respectively; the coefficients of correlation are probably slightly less. Such low coefficients must mean a relatively very high effect of circumstance, and in such a case *the initial correlation between the daughters and grand-daughters of one mother must be exceedingly small* and the approach towards the ultimate value extremely slow. I have calculated the figures given below as a rough illustration of the sort of thing to be expected; the figures were obtained on the assumption of asexual propagation, which is not quite the same (on the continuous-variation hypothesis) as self-fertilization. The slowness of increase in the coefficient of correlation is possibly greater than that to be expected in Professor Johannsen's cases, as the ultimate coefficient (·117) is a little lower, but one cannot say for certain, not knowing the grand parental coefficient.

Table illustrating the gradual increase in the intensity of heredity between the successive generations of a pure line on the assumption of continuity of variation in the germ-plasm. The ancestor is reckoned as generation 0, and $r_{1\cdot 2}$, $r_{2\cdot 3}$, etc., are the coefficients of correlation between individuals of generation 1 and individuals of generation 2, between the latter and individuals of generation 3 and so on. s_1, s_2, s_3, etc. are the relative standard deviations of the successive generations.

$r_{1\cdot 2}$	·0072	s_1	1000	$r_{11\cdot 12}$	·0656	s_{11}	1032
$r_{2\cdot 3}$	0143	s_2	1004	$r_{12\cdot 13}$	·0698	s_{12}	1034
$r_{3\cdot 4}$	·0211	s_3	1007	$r_{13\cdot 14}$	·0738	s_{13}	1037
$r_{4\cdot 5}$	·0276	s_4	1010	$r_{14\cdot 15}$	·0774	s_{14}	1039
$r_{5\cdot 6}$	·0340	s_5	1014	$r_{15\cdot 16}$	·0809	s_{15}	1041
$r_{6\cdot 7}$	·0400	s_6	1017	$r_{16\cdot 17}$	·0840	s_{16}	1043
$r_{7\cdot 8}$	·0457	s_7	1020	$r_{17\cdot 18}$	·0869	s_{17}	1045
$r_{8\cdot 9}$	·0511	s_8	1024	$r_{18\cdot 19}$	·0896	s_{18}	1046
$r_{9\cdot 10}$	·0563	s_9	1026	$r_{19\cdot 20}$	·0920	s_{19}	1048
$r_{10\cdot 11}$	·0611	s_{10}	1029	$r_{20\cdot 21}$	·0943	s_{20}	1049
	ultimate value of r				·1172		
	ultimate value of s				1064		

Now the probable error of any correlation coefficient less than 0·2 or so is $\pm$·021 for a thousand observations, and the coefficient only reaches this value between the third and fourth generations. That is to say, if a thousand individuals (single beans) were gathered at each generation the correlation would only begin to be fairly well marked when the fourth generation of the "pure line" had been harvested. In many of Professor Johannsen's "pure lines" there were only two or three hundred seeds weighed; seven or eight generations might be grown before such numbers gave a clear result. The large effect of *definite* circumstances on plants (producing fluctuations not of the nature of errors of sampling at all) would further tend to obscure the results. The increase in variation, it should be noted would also be difficult to attest. The whole increase in twenty generations is only 5%, and the probably error of a standard deviation (based on 1000 observations) is approximately 2%.

So far as they have gone then, Professor Johannsen's results do not seem in any way to contradict but rather to support the hypothesis of continuous variation. The results of succeeding generations will be awaited with a good deal of curiosity, if, as is to be hoped, the breeding of the "pure lines" is continued in such a way as to enable the above view as to the gradual approach towards the behaviour of an ordinary race to be tested; whether the view be confirmed or disproved it is equally desirable to have the experimental evidence.

Professor Johannsen's work is certainly one of the most important contributions to the theory of heredity of recent years, and his results should be studied and judged in the original by all who are interested in the subject. The mode of treatment is novel, and the study of "pure lines" a thoroughly sound procedure well calculated to elucidate the nature of intraracial heredity. One may add that the refreshing width of sympathy and sobriety of the author's style contribute not inconsiderably to the pleasure of the reader.

21

Reprinted from pages 1–4 and 32–34 of *Selection and Cross-breeding in Relation to the Inheritance of Coat-pigments and Coat-patterns in Rats and Guinea-pigs*, Carnegie Institution of Washington Publication No. 70, 1907

SELECTION AND CROSS-BREEDING IN RELATION TO THE INHERITANCE OF COAT-PIGMENTS AND COAT-PATTERNS IN RATS AND GUINEA-PIGS.

By Hansford MacCurdy and W. E. Castle.

CONTINUOUS VERSUS DISCONTINUOUS VARIATIONS AS FACTORS IN EVOLUTION.

It is generally agreed that the course of evolution is largely influenced by two factors, variation and heredity; but opinions differ as to what sorts of variation have evolutionary significance and as to the manner of their inheritance.

It has been recognized by several investigators that variations are of two distinct sorts. Bateson has called these two sorts of variation continuous and discontinuous; more recently De Vries has called them fluctuations and mutations, respectively.

By continuous variation (or fluctuation) we understand ordinary individual variation within a species. The individuals differ among themselves in size, color, and other structural features. By examining a considerable number of them we can form an idea of what is the commonest (or *modal*) condition as regards each structural feature; and likewise what is the average (or *mean*) condition.

Usually, but not always, the modal and mean conditions are approximately the same, and any other condition is the less frequent in occurrence, the greater its deviation from them. It follows that the most extreme condition observed is connected with the most usual (or modal) condition by an unbroken series of intermediate conditions, and we may call the series as a whole "continuous." The distribution of the individuals in such a series is governed by the laws of "chance," and may be successfully analyzed by statistical methods.

We commonly think of a "chance" result as something entirely beyond the control of law, but in reality such is not the case. Nothing is beyond the control of law. If a blindfolded person puts his hand into an urn containing a mixture of black and of white balls, it is a matter of chance whether he grasps a black or a white ball; but if he repeats the operation a considerable number of times, it is perfectly certain that he will draw balls of both sorts in approximately the same proportions in which they occur in the jar. The result is a "chance" one, but controlled by a perfectly definite mathematical law.

A "chance result" has been aptly defined as the result of a number of causes acting independently of each other. If this is a valid definition, then a continuous series of variations is due to no single cause but to several mutually independent ones. Some of the causes may be external in origin, others internal; some temporary in their action, others permanent. It should not surprise us, therefore, to find that continuous variations differ greatly in the degree of their inheritance. De Vries, indeed, has maintained that they are not inherited at all, except temporarily; that selection of abmodal variations from a continuous series is unable permanently to modify a race; that the modifications will persist only so long as selection continues, but will speedily disappear when selection is arrested. This conclusion, however, seems to us altogether too sweeping. *A priori* there is no reason to suppose that *all* the causes operative to produce continuous variation are external in origin and temporary in action, as De Vries's conclusion would seem to imply. If there are in operation, in the production of a continuous series of variations, causes internal in origin, resident in the constitution of the germinal substance, so much of the result as is due to those causes should be inherited and so should be permanent. De Vries, we believe, has overlooked this factor entering into the problem. He has assumed that all the causes of continuous variation ("fluctuations") are either external in origin or due to conditions of the germinal substance purely temporary. He holds, we believe rightly, that all inheritance is due to germinal modification; but assumes, we believe without sufficient warrant, that permanent germinal modification is not a factor in the production of fluctuations.

Another category of variations, discontinuous variations (which include the mutations of De Vries), is considered by Bateson and De Vries as the true and only expression of permanent germinal modification. But, granting the truly germinal origin of mutations, it does not follow that they are the *only* product of germinal modification.

A discontinuous variation, as the name suggests, is unconnected by intermediate conditions with the usual (*modal*) condition of the species. It represents a change, more or less abrupt, from the modal condition of the species, and is strongly inherited, a fact which indicates clearly its exclusively germinal origin.

In the category of discontinuous variations belong abrupt changes in pigmentation and hairiness among both animals and plants, changes in the number of digits or of the number of phalanges in a digit among vertebrates, in the presence or absence of horns among animals and spines among plants, and other similar conditions.

Such changes are not the result of selection; they often appear, as it seems, spontaneously, and they are permanent in the race, if isolated.

De Vries maintains that all species-forming variations are of this sort; that selection is unable to form new species, because it can neither call into existence mutations nor permanently modify a race by cumulation of abmodal fluctuations. Darwin, on the other hand, and the great majority of his followers, while admitting that races are occasionally produced by discontinuous or "sport" variation, ascribe evolutionary progress chiefly to the cumulation through long periods of time of slight individual differences, such as De Vries calls fluctuations. The issue between the two views is sharp and clear. According to De Vries, if we rightly understand him, selection is not a factor in the *production* of new species, but only in their *perpetuation*, since it determines merely what species shall survive; according to the Darwinian view, new species arise through the direct agency of selection, which leads to the cumulation of fluctuating variations of a particular sort.

De Vries and the Darwinians differ not only as to the part which selection plays in evolution, but also as to the nature of the material upon which selection acts. According to De Vries, species are not modified by selection; mutations *are* new species and selection determines only what mutations shall survive, fluctuations having no evolutionary significance. On the Darwinian view, all species, whether arising by mutation or not, are subject to modification by selection.

A great deal can be said in favor of each of these contrasted views, but discussion is at present less needed than experimental tests of the views outlined. To De Vries we owe much for showing that such tests are possible.

It was our purpose to make tests of this sort when we undertook the experiments described in this paper. The questions to which principally attention has been directed are these: (1) Can discontinuous variations be modified by selection alone? (2) Can discontinuous variations be modified by cross-breeding? A negative answer to these questions will support the view of De Vries; an affirmative answer will support the Darwinian view, because it will show that through selection new conditions of organic stability can be obtained; that is, new species may be produced.

The material used consisted of certain discontinuous variations in the color-pattern of rats. The general result obtained is this: Various color-patterns, like the several pigments found in the rodent coat, are mutually alternative in heredity. Each group of individuals referred to the same type of color-pattern forms a continuous series fluctuating in accordance with the laws of chance about a common modal condition. The different types in general do not overlap; they form a discontinuous series. Now, these types may be modified in two different ways: (*a*) By selection of abmodal variates within the same continuous series, and (*b*) by cross-breeding between different types. There is no evidence that one of these methods has effects less permanent than the other. So far, then, as these experiments go, they support the Darwinian view rather than that of De Vries.

[*Editor's Note:* Material has been omitted at this point.]

CONCLUSIONS.

The results of selection brought to bear upon the coat-pattern are seemingly very different in rats and in guinea-pigs, yet a careful analysis of the facts shows the results to be not so dissimilar in the two cases as they at first thought appear.

In both rats and guinea-pigs we can by selection increase or decrease at will the average extent of the pigmented areas. In both rats and guinea-pigs the extent of the pigmented areas varies continuously, and out of these continuous variations permanent modification of the pigmentation can be secured.

Reduction in the total amount of the pigmentation is attended in rats by restriction of the pigment to very definite areas, whereas in guinea-pigs it may be distributed in any or in all of a series of spots. Herein lies the whole difference between the two cases. When in rats we select for reduced pigmentation, we get animals with a narrow or interrupted back-stripe and with a less extensive hood; when in guinea-pigs we make a similar selection, we get animals with fewer or less extensive spots. We can not in guinea-pigs decide arbitrarily *which* areas shall be pigmented (except, possibly, in the case of nose spots), any more than in rats we can at the same time increase the extent of the hood and decrease that of the back-stripe.

In rats, we have as a result of pigment reduction a series of coat-patterns, each breeding true within certain limits; in guinea-pigs, the fluctuation in the extent of the pigmented areas is probably no greater than in rats, but because the pigmented areas do not disappear in as definite an order during pigment reduction, we have no constant coat-patterns. Nevertheless there is every reason to suppose that different degrees of pigmentation are inherited in Mendelian fashion in guinea-pigs, precisely as they are in rats. If the pigment reduction followed a definite course in guinea-pigs, as it does in rats, this would be easily recognizable in the coat-pattern. As it is, measurement of the extent of the pigmented areas would be necessary to make it apparent. This we have not undertaken to do in the case of guinea-pigs; we have merely taken account of the regions pigmented, not of their extent. This probably explains in part why regression is observed in the selection experiments with guinea-pigs, but not with uniformity in those with rats. In guinea-pigs we attempted by selection to restrict the *number* of the pigmented areas; this was found to be impossible except as it occurred incidentally to reduction in the total *amount* of pigmentation. The regression occurred in *number* of pigmented areas, not, so far as we know, in the total amount of the pigmentation. We have no doubt, however, that such regression would be found to occur in cases in which extreme variates were selected. We have found it so in selecting black-eyed white guinea-pigs, those with no pigment except in the eye. Almost invariably the young of such animals have borne more pigment than did their parents. A similar result would doubtless follow selection of self-pigmented rats obtained from Irish parents. No doubt many of the young would bear some white fur. With selection of less extreme variates, regression less extreme may possibly occur, though our statistical observations do not show any regression in the case of rats.*

If regression does occur, can we with propriety consider the effects of selection permanent? De Vries has answered this question in the negative on the basis of his selection experiments with maize, striped flowers, double buttercups, and other similar material. It seems to us, however, that the answer should be qualified. The final result will depend upon the amount and the persistency of the regression. In De Vries's experiments with maize, as in those of Fritz Müller ('86), the regression grows less with each selection. If this continued, the regression should ultimately become a negligible quantity. After repeated selection for a desired extreme condition, the race should become stable at a condition only a little less extreme than that selected.

De Vries's fine series of selection experiments with the buttercup (*Ranunculus bulbosus*) seems to the writers scarcely to justify the conclusion that selection has no permanent effects. Starting with a one-sided or "half-Galton" variation curve, with a range from the modal number, 5, upward to 13, De Vries was able by selection for an increased number of petals to raise the mode to 11, the average to 8.6, and the upper limit of variation to 31, and to obtain a two-sided, or Galtonian, variation curve with only a moderate amount of skewness, and with greatly diminished regression. All this was accomplished within five generations.

We consider the selection question still an open one. Further experiments and longer continued ones are needed. Our own observations, so far as they go, and those of Fritz Müller and De Vries, lead us to think that selection is a most important factor, not only in the isolation of discontinuous variations, but also in their production.

Further, we are far from convinced that all evolutionary progress is to be attributed to discontinuous variations, any more than to Mendelian inheritance. The distinction between continuous and discontinuous variations is a useful one, just as that between alternative and blending inheritance, but a sharp line of division can be drawn in neither case. The hooded and Irish coat-patterns of rats are recognized discontinuous variations, alternative in inheritance, yet our lot *M* of hooded rats is as nearly intermediate between typical hooded and typical Irish rats as anything that can well be imagined. The coat-patterns of fancy rats, though discontinuous as they ordinarily occur, can be transformed into continuous variations. Concerning

*April, 1907. In this year's experiments we are getting some evidence of the expected regression.

Müller, F., 1886, Ein Zuchtungsversuch an Mais, *Kosmos* **19**:22–26.

the hooded and Irish patterns, Doncaster (:06), after an extended experience, says (p. 216): "Only once have I had any hesitation in classing a rat as belonging to one or the other." Yet we have seen that by cross-breeding and selection these same discontinuous groups can be made continuous by the production of any desired number of intermediate groups, each varying continuously about a different mode.

Again, though the inheritance is clearly Mendelian, when hooded and Irish rats are crossed, the gametes formed by cross-breds are not pure, but modified, each extracted pattern being changed somewhat in the direction of that pattern with which it was associated in the cross-bred parent. This means simply that the inheritance, though in the main alternative, is to some extent blending.

Since it is impossible to make a sharp distinction between continuous and discontinuous variations, as well as between blending and alternative inheritance, it is fallacious to assign all evolutionary progress to one sort of variation or to one sort of inheritance.

TABLES.

TABLE 1.—*Actually different classes into which we may expect the five visibly different classes of albino and partial-albino rats to fall.*

[Twenty-two of the twenty-seven classes enumerated have been proved to exist.]

	Classes visibly different.				
	Gray Irish.	Gray hooded.	Black Irish.	Black hooded.	Albino.
Classes expected	1 GI	1 GH	1 BI	1 BH	1 W[BH]
	2 GI(W)	2 GH(W)	2 BI(W)	2 BH(W)	1 W[BI]
	2 GI(B)	2 GH(B)	2 BI(H)	...	1 W[GH]
	2 GI(H)	4 GH(W.B)	4 BI(W.H)	...	1 W[GI]
	4 GI(W.B)	...	...	...	2 W[B-IH]
	4 GI(W.H)	...	...	...	2 W[G-IH]
	4 GI(BH)	...	...	...	2 W[GB-H]
	8 GI(W.BH)	...	...	...	2 W[GB-I]
	...	...	...	...	4 W[GB-IH]
Total...	27	9	9	3	16

B, black; G, gray; H, hooded; I, Irish; W, total albinism. Symbols indicating unseen recessive characters are placed within parentheses ( ), those indicating characters latent in albinos are placed within brackets [ ].

When two symbols only are used the first refers to color, the second to coat-pattern.

When more than two symbols stand together within brackets, those which refer to color are placed at the left of a hyphen, those referring to pattern at its right.

Total albinism is indicated by W. When albinism is recessive with other characters in pigmented individuals, the W will be separated by a period from the symbols designating the other characters.

The numerals prefixed to the several class-designations indicate the expected frequencies of the classes when individuals are mated *inter se* which have the characters indicated by the designation GI (W.BH).

[*Editor's Note:* Material has been omitted at this point.]

Doncaster, L., 1906, On the Inheritance of Coat-Colour in Rats, *Proc. Cambridge Phil. Soc.* **13**:215–227.

22

Reprinted from *Z. Induk. Abstamm. Vererbungsl.* **2**:257–275 (1909)

Is there a Cumulative Effect of Selection?

Data from the Study of Fecundity in the Domestic Fowl[1])

By **Raymond Pearl,** Ph. D. and **Frank M. Surface,** Ph. D.

Introduction.

The purpose of this paper is to present briefly certain results of a series of selection experiments[2]) which has been carried on at the Maine Experiment Station with poultry during the past eleven years having as its purpose to get data on the question which stands as the title of this paper. The specific character which was the object of selection in these experiments was "egg production" or fecundity. This is a character which is particularly suited to selection studies because it is expressed in integral units and on that account it is possible to observe with much exactness and precision the effect of any selective or other experimental procedures.

Before undertaking specifically the discussion of our results it will be well to review very briefly the history of biological opinion regarding the question of whether there is a definitely cumulative effect of selection. It has long been held as one of the most fundamental principles of practical breeding, both with animals and plants, that the surest way in which to bring about improvement in a strain was to select for breeding the individuals of superior quality. Certainly until comparatively recently a great many breeders would

---

[1]) Papers from the Biological Laboratory of the Maine Agricultural Experiment Station. Orono, Maine, U. S. A. No. 12.

[2]) The detailed report oi the results of these experiments will be found in the following publications by the present writers:

(a) A Biometrical Study of Egg Production in the Domestic Fowl. Part I — Variation in Annual Egg Production. Bur. of Animal Industry, U. S. Dept. of Agriculture, Bulletin 110, 1909.

(b) Data on the Inheritance of Fecundity Obtained from the Records of Egg Production of the Daughters of "200-Egg" Hens. Maine Agricultural Experiment Station Bulletin 166, pp. 49—84, 1909.

have said that this was the only absolutely certain way to insure improvement. It is merely a statement of fact to say that even now the vast majority of practical breeders hold to this opinion. It is embodied in many maxims of practical breeding such as "breed from the best to get the best", "like produces like", etc.

Widespread as it has always been among practical breeders this view as to the effect of selection can hardly be said to have acquired a definite scientific status until after the publication of the "Origin of Species". A great deal of the material which Darwin used in this work and also in the "Variation of Animals and Plants under Domestication" was obtained from the records and published statements of practical agricultural breeders. Darwin himself was firmly convinced of the general truth of the contention that improvement in domesticated animals had been brought about chiefly by the continued selection of small, favorable variations. This is sufficiently shown by the following statement: "We cannot suppose that all the breeds were suddenly produced as perfect and as useful as we now see them; indeed, in many cases we know that this has not been their history. The key is man's power of accumulative selection: nature gives successive variations; man adds them up in certain directions useful to him. In this sense he may be said to have made for himself useful breeds."[1]) Still more definite is the following: "If selection consisted merely in separating some very distinct variety, and breeding from it, the principle would be so obvious as hardly to be worth notice; but its importance consists in the great effect produced by the accumulation in one direction, during successive generations, of differences absolutely inappreciable by an uneducated eye — differences which I for one have vainly attempted to appreciate."[2])

The view that there was in practically all cases of the improvement of domesticated plants and animals by breeding a definitely cumulative effect of selection may be said to have held the field almost undisputed until the publication of de Vries' "Mutationstheorie." This investigator's searching critique of the data furnished by the results of agricultural breeding work showed clearly enough that certainly all the improvement which has taken place in plants and animals under domestication can not be explained as a cumulative result of the continued selection of favorable variations.

---

1) Origin of Species. Chapter I, p. 25.

2) Loc. cit. p. 26.

During the last few years there has been a considerable amount of careful and critical experimental work done to test the method of action and limitations of continued selection in breeding. In particular reference should be made in this connection to the work of the Svalöf Experiment Station in practical plant breding;[1]) to the experimental work of Johannsen[2]) with beans, and to the extensive and thorough investigations of Jennings[3]) on selection in Paramecium. The broad general result reached by these workers may be fairly stated as follows: From a mixed "general" population it is possible by a single selection to isolate pure strains ("pure lines," "homozygote strains," "pure races") which will breed true and not revert to the mean of the general population from which they were isolated, regardless of whether further selection is practiced or not. It is impossible to demonstrate any cumulative effect of continued selection within the pure strain. Continued breeding from the extreme individuals of such a pure strain ("fluctuating" variants) does not change the mean of that strain. From these considerations it follows that it will be difficult or impossible to make any definite and permanent change in the mean of a general population simply and solely by continued selection of extreme individuals, because in the vast majority of cases such individuals will be extreme fluctuating variants rather than mutants. Correns[4]) says (loc. cit. p. 51): "Die Zuchtwahl, die künstliche sowohl wie die natürliche, hat, auf die individuellen Varianten angewandt, jedenfalls keinen bleibenden Erfolg, wahrscheinlich gar keinen." Jennings, in concluding his paper, makes a statement of similar import. He says (loc. cit. p. 522): "Certainly, therefore, until some one can show that selection is effective within

---

[1]) For summaries of this work in languages more generally read than Swedish cf.: (a) Ulander, A., Die schwedische Pflanzenzüchtung zu Svalöf. Eine kurze Darstellung. Jour. f. Landwirtsch. Bd. 54, 1906, pp. 105—124. Pl. II—VII (b) Constantin, J. Le transformisme appliqué a l'agriculture. Paris (Alcan) 1906. Chap. IX. (Le Laboratoire de Svalöf et la Mutation), pp. 71—96. (c) De Vries, H. Plant Breeding. Chicago (Open Court) 1907, Chap. II (The Discovery of the Elementary Species of Agricultural Plants by Hjalmar Nilsson) pp. 29—106.

[2]) Johannsen, W. Über Erblichkeit in Populationen und in reinen Linien. Jena (Fischer) 1903. Pp. 68.

[3]) Jennings, H. S. Heredity, Variation and Evolution in Protozoa. II. Heredity and Variation of Size and Form in Paramecium with Studies of Growth, Environmental Action and Selection. Proc. Amer. Phil. Soc. Vol. XVI, pp. 393—546. 1908.

[4]) Correns, C. Experimentelle Untersuchungen über die Entstehung der Arten auf botanischem Gebiet. Arch. rass. u. gesellsch. Biol. Bd. I, pp. 27—52. 1904.

pure lines, it is only a statement of fact to say that all the experimental evidence we have is against this."

The bulk of the evidence which has developed up to the present time in favor of the view that there is not a definitely accumulative effect following the continued selection of fluctuating variations has been obtained from experiments on plants. There have been practically no systematic investigations of the effect of selection continued through many generations on higher animals, but it obviously is a matter of the greatest importance, both theoretically and practically, to determine in how far the principles which have been found to hold with reference to selection in plants also apply with reference to animals. The thorough investigation of the selection problem by Jennings referred to was, to be sure, carried out on an animal form, Paramecium, The method of reproduction in this protozoan form (fission) is. however, so different from sexual reproduction in vertebrates, that it must make one cautious (as Jennings himself is) about generalizing too extensively from these results. What is needed is a special investigation of the selection in a higher vertebrate. By preference this should be a domestic animal, because of the obviously great practical importance of questions involved. So far as we are aware the investigations which form the subject of the present paper, constitute the first experimental study of the effect of selection in a higher animal to be carried out with large numbers of individuals through a number of generations and with exact numerical records of the character studied in each individual.

Plan of the Experiments.

It is our purpose to discuss the results of two distinct and separate but supplementary experiments. These may be designated as follows:

I. Experiment in continued selection of fluctuating variations in fecundity.

II. Experiment regarding the inheritance of fecundity.

I. Experiment in continued selection of fluctuating variations in fecundity. In 1898 there was begun at the Maine Agricultural Experiment Station an experiment to determine whether egg production in the domestic fowl could be increased by the continued selection of the highest egg producers as breeders. This experiment was planned and started by Director C. D. Woods and the late Professor G. M. Gowell. An exact record was made of the egg production of each hen during the first year of her life; trap nests being used to furnish the individual records. In 1907 the records of egg production which had accumulated from the beginning of the experiment up to that time, were turned over to the present writers for analysis. The present paper sets forth certain of the results of the analysis made. The plan of the experiment begun in 1898 was to make from a then superior strain of Barred Plymouth Rock hens, which had been pure bred for a long time by Professor Gowell, a continuous close and intense selection with reference to egg production. The practice in breeding was to use as mothers of the stock bred in any year only hens which laid between November 1 of the year in which they were hatched and November 1 of the following year, 160 or more eggs. After the first year, all male birds used in the breeding were the sons of mothers whose production in their first laying year was 200 eggs or more. Since the normal average annual egg production of these birds may be taken to be about 125 eggs, it will be seen that the selection practiced was fairly stringent.

Close inbreeding was not practiced. It was always possible to avoid this, since after the first few years of the experiment the flocks were very large (always containing more than 500 birds and usually nearer a thousand). While there was no close inbreeding no "new blood" was introduced into the strain from the outside during the period of the experiment.

While every effort was made to preserve uniform environmental conditions during the course of the experiment certain unavoidable environmental accidents occurred in certain of the laying years. These accidents may be held to have affected the egg production in those years adversely. Therefore, in discussing the results it will be necessary to make certain corrections for them. For a detailed account of their nature as well as for further details regarding the conduct of the experiment the more extensive report (loc. cit.) must be consulted.

II. Experiment regarding the inheritance of fecundity. In 1907 the experiment described above, having led to definite results was brought to an end. There was planned for 1908 a new experiment designed to test from another standpoint the conclusions which had been tentatively reached from the earlier experiment. In the conducting of the long selection experiment the females used as breeders were grouped into two classes, viz., (a) "unregistered" or birds laying 160 to 199 eggs in the pullet year, and (b) "registered" or birds laying 200 or more eggs in the pullet year.

It had been noted that the daughters of the socalled "registered" hens (namely hens that had produced 200 or more eggs each in the pullet year) did not usually make high egg records. The "200-egg" birds which made up the "registered" flock came, in most instances, from the "unregistered" mothers.

Experiment II was planned to answer the two following questions:

1. Will the daughters of high laying hens ("200-egg" birds) on the average produce more eggs in a given time unit than will birds of less closely selected ancestry?

2. What data do the performance records of such selected birds afford regarding the inheritance of fluctuating variations in egg producing ability in the domestic fowl?

The experiment was carried out according to the following plan: On the first of November, 1907, there were put into House No. 2, of the Station plant, 250 pullets. Each of these was the daughter of a hen that had laid approximately 200 eggs in her pullet year. These 250 pullets were divided into flocks of 50 each and were fed and handled in every way exactly alike. At the same time that these 250 "registered" pullets (so-called because from "registered" mothers), were put into the house there were also put in 600 other Barred Plymouth Rock pullets. These other pullets were of approximately the same age as the 250 "registered" pullets and differed in their breeding only in respect to their mothers. They came from hens that had laid less than 200 eggs during the pullet year and more than 160. "Registered" cockerels (from the "200-egg line") were used as the male parents for all the pullets both "registered" and "unregistered". The 600 "unregistered" birds were divided into flocks as follows: Two flocks of 50 birds each were kept in two pens in House No. 2, exactly like the pens in which the "registered" birds were kept. The remaining 500 birds were divided into four flocks — two of 100 birds each and two of 150 birds each and housed in the four pens of House No. 3. These pens are essentially like those of House No. 2, differing chiefly in the matter of size. A trap nest[1]) record was kept of the exact individual egg production of each of there birds.

---

[1]) For a description of the trap nest used in the breeding work of the Station see "Appliances and Methods for Pedigree Poultry Breeding" by R. Pearl and F. M. Surface. Me. Agric. Exp. Station. Bulletin No. 159, pages 239—274. 1908.

Chief Results of Selection Experiment I.

In considering the results of the long selection experiment I the the three following questions present themselves as of first importance.

1. What is the general character of the variation exhibited in first-year egg production (fecundity) of the domestic fowl? Do we have here typical fluctuating variability?

2. Did the mean or average annual egg production per bird increase or decrease or remain unchanged during the course of the intensive and long continued selection practised in this experiment?

3. Was there any definite change in the variability of the flocks in respect to egg production during the course of the experiment? Did the birds come truer to type in respect to egg production at the end of the experiment than they did at the beginning?

Variation in Annual Egg Production. A careful biometrical analysis of variation in annual egg production (fecundity) was made with our data. The methods of Pearson were used in this analysis. Without going into detail regarding the results of this part of the work it may be said that the data warrant the following conclusions:

(a) Variation in egg production (fecundity) in the domestic fowl is, so far as can be determined, continuous and "fluctuating". It is of such character as to give rise to unimodal frequency curves, similar to those known for variation in other characters and organisms.

(b) The observed frequency polygons were all capable of satisfactory graduation by one or another type of Pearson's generalized probability curves. The values of the analytical constants obtained indicate that, in general, these poultry fecundity data demand limited range curves for their graduation. The majority of the curves are of Pearson's Type I (Skew curve of limited range).

c) The skewness or asymmetry of these variation curves is, generally speaking, not great, though in most cases significant in comparison with its probable error.

The most significant point for our present purpose is that in egg production we are dealing with a character which varies strictly in the usual or normal "fluctuating" continuous manner.

The general character of these variation curves may be illustrated by a single example. Figure 1 gives the frequency polygon and fitted curve for variation in egg production exhibited by 275 birds kept in flocks of 150 birds each in the laying year 1905—06.

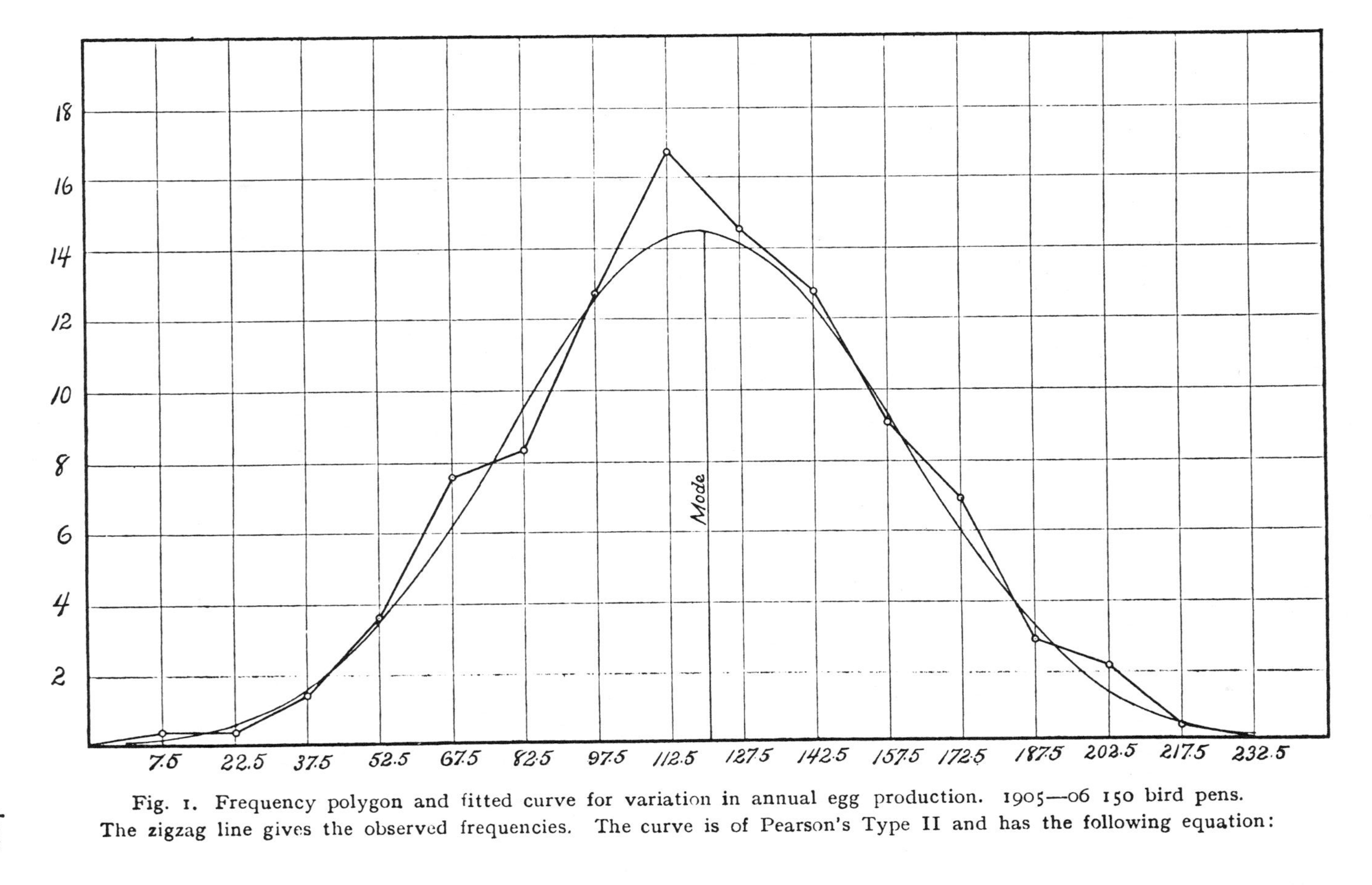

Fig. 1. Frequency polygon and fitted curve for variation in annual egg production. 1905—06 150 bird pens. The zigzag line gives the observed frequencies. The curve is of Pearson's Type II and has the following equation:

$$y = 39.86 \left(1 - \frac{x^2}{132.1586}\right) 8.8636.$$

Results of continued selection for high egg production. The essential data giving the results of the selection carried out in this experiment are shown in Table I. This table gives for each year of the experiment the following data: (a) the actual mean or average egg production, (b) the modified average production, (c) the variability in regard to egg production as measured by the coefficient of variation, and (d) the number of birds included.

Table I.

Results of Selection Experiment I.

Laying Year	Observed Mean Egg Production	Modified Mean Egg Production	Coefficient of Variation	Number of birds furnishing data
1899—1900	136.36 ± 3.55	136.36	32.29 ± 2.02	70
1900—1901	143.44 ± 3.29	143.44	31.38 ± 1.78	85
1901—1902	155.58 ± 3.79	155.58	25.03 ± 1.83	48
1902—1903	136.48 ± 2.19	159.15	28.82 ± 1.22	147
1903—1904	117.87 ± 1.75	129.14	35.14 ± 1.17	254
1904—1905[1]	134.60 ± 1.98	134.07	36.77 ± 1.17	283
1905—1906[1]	140.31 ± 1.81	154.09	25.48 ± 0.97	178
1906—1907[1]	114.16 ± 1.74	142.77	30.91 ± 1.18	187

The data of Table I are shown graphically in figs. 2 and 3. Figure 2 shows the change in the observed and modified averages, and the best fitting straight lines to these data, as determined by the method of least squares, Figure 3 shows the change in the variability of the flocks (coefficient of variation), the observational data again being smoothed by a straight line fitted by the method of least squares.

Observed means $y = 148.48 - 3.10\ x$
Modified ,, $y = 144.13 + 0.043\ x$
Variability $y = 30.56 + 0.039\ x.$

where x denotes the year.

From the data of Table I and Figures 2 and 3 we note the following points:

[1]) Data for birds kept in flocks of 50 birds each used in these years. The reasons for this procedure are given in the detailed report q. v.

1. There was no definite increase in mean annual egg production during the period of eight years covered by the statistics. As is to be expected, there are up and down fluctuations in the several years, but the general trend of the means, as indicated by the fitted straight lines in Fig. 2, is certainly not upward. Taking the actual means the trend of the line is clearly downward. The line for the modified

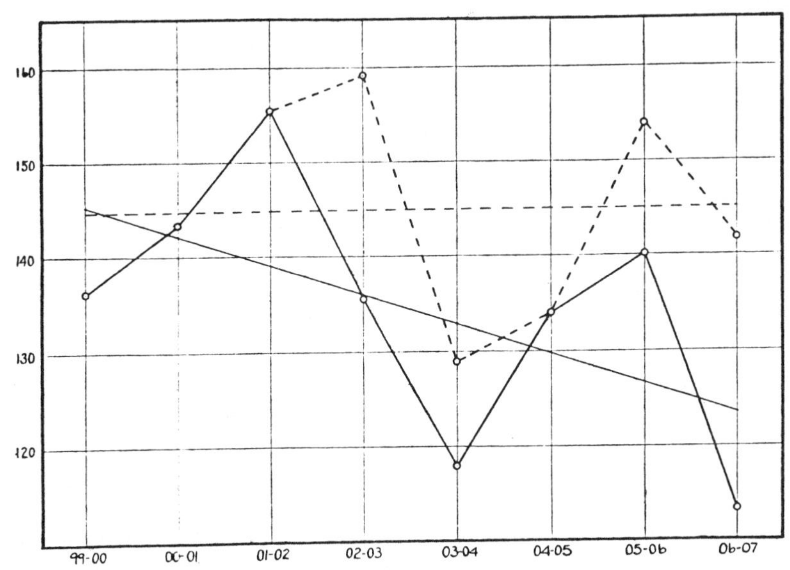

Fig. 2. Diagram showing change in mean egg production during the progress of the selection experiment. The unbroken lines give the observed means, and the dotted lines the modified means.

means is horizontal. There is no evidence whatever of any cumulative effect of the selection practised.

2. The general trend of the variability as shown by the fitted line in Figure 3, is slightly to increase during the course of the experiment. No stress is to be laid on this slight increase. The important thing is that there is no evidence whatever of any decrease in relative variability as a result of the long continued and intensive selection.

Chief Results of Selection Experiment II.

The plan and purpose of this experiment have been stated above (p. 262) and need not be here repeated. We may first consider the the question as to what evidence the experiment furnished regarding the inheritance of fluctuating variations in egg production. The comparative figures for the egg production of mothers and daughters are given in Table II. In discussing the results of this experiment the egg production of the winter months (November 1 to March 1) and of the spring months (March 1 to June 1, the natural mating and breeding season of the birds) will be considered separately.

From the data set forth in this table and additional ones given in the detailed report the conclusions given below may be drawn. It will be understood that considerations of space forbid giving here the entire evidence on which these conclusions rest. For the detailed

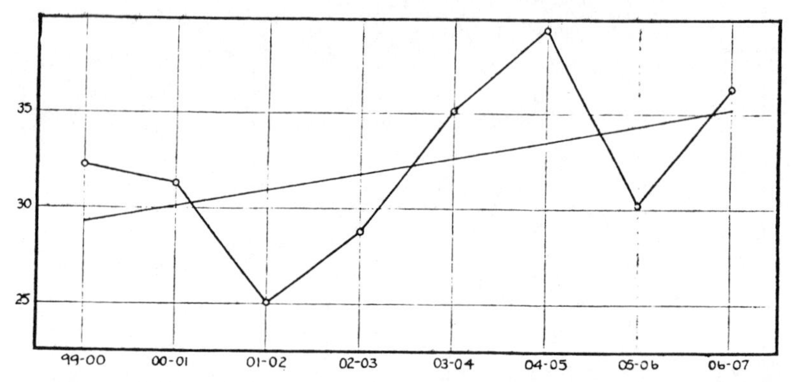

Fig. 3. Diagram showing change in variability of flocks in egg production during the course of the selection experiment. The zigzag line gives the observed coefficients of variation.

presentation and critical analysis of this evidence the complete papers already referred to (p. 257 supra) must be consulted.

1. The daughters of "200-egg" or "registered" hens are markedly inferior to their mothers in egg production for both of the periods studied. This difference is particularly great in the winter (Nov. 1 — March 1) period. It is evident that the daughters do not belong in anything like the same class as the mothers as winter egg producers. The mothers' average production for the corresponding period of their pullet year was nearly 4 times as great as the daughters'. (Exactly $45.5 \div 12.7 = 3.7$.) This great reduction of the daughters' average winter production below the mothers' is most striking and unexpected. It is to be expected on general grounds that there would be some regression, but so much as this would hardly have been anticipated.

2. Quite apart from the question of the average production of mothers and daughters the data gives no indication that there is any sensible correlation between individual mothers and individual daughters

Table II.

Showing the Egg Records of "Registered" Hens and the Number and Egg Records of Their Daughters.

Band number of "registered mother" hen	Mother's egg production from Nov. 1 of pullet year to Nov. 1 of following year	Number of daughters of each "registered" hen in the experiment			Winter egg production Nov. 1—Mar. 1		Spring egg production Mar. 1—June 1	
		Total	Nov. 1—Mar. 1	Mar. 1—June 1	Mother's	Daughter's average	Mother's	Daughter's average
7	193	12	12	12	33	14.83	81	47.75
33	192	8	8	8	30	13.37	68	53.25
578	196	4	4	4	42	23.75	59	42.25
253	200	9	8	7	65	23.87	45	38.14
46	184	3	3	3	64	15.33	56	51.00
460	208	8	8	8	61	15.50	60	44.87
617	183	4	4	4	47	15.25	56	38.00
42	216	7	7	6	46	13.28	67	52.83
169	210	17	15	15	70	13.47	54	44.43
150	193	5	5	5	56	18.00	53	50.80
236	180	7	6	5	24	20.83	62	46.20
152	200	6	6	6	49	18.83	69	43.33
105	217	8	8	8	68	17.25	66	48.12
386	203	9	9	9	55	4.88	59	40.88
911	196	3	3	3	52	27.33	62	55.33
510	212	3	3	2	69	0.00	59	12.50
404	203	2	2	2	55	19.50	56	62.00
174	208	6	6	6	43	24.00	68	44.16
166	221	7	7	7	66	7.14	56	46.57
2	190	10	10	10	48	17.90	60	45.80
505	204	6	6	5	58	17.66	43	38.00
464	202	7	7	7	77	19.57	62	51.28
32	203	9	9	9	64	11.00	55	50.33
111	?	10	—	—	—	—	—	—
379	203	4	4	4	63	24.25	66	47.50
130	208	3	3	3	68	14.00	51	53.00
49	200	4	4	3	29	18.00	65	42.33
351	201	5	5	5	68	15.20	49	53.80
9	186	8	7	7	60	14.14	49	59.85
614	?	9	—	—	—	—	—	—
14	197	6	5	4	66	2.40	68	47.75
349	203	4	4	3	51	10.25	55	37.33
303	246	4	4	4	83	7.25	60	29.70
Total number of "mothers" represented	Average "mother's" egg[1]) production	Totals			Mother's average Nov.—Mar. production	Daughter's average, Nov.—Mar., production	Mother's average Mar.—June, production	Daughter's average Mar.—June, production
33	201.8	217	192	184	55.80	15.29	59.13	46.61

[1]) Omitting the two birds without records, No. 111 and No. 614.

in respect to egg production. It is quite conceivable that the

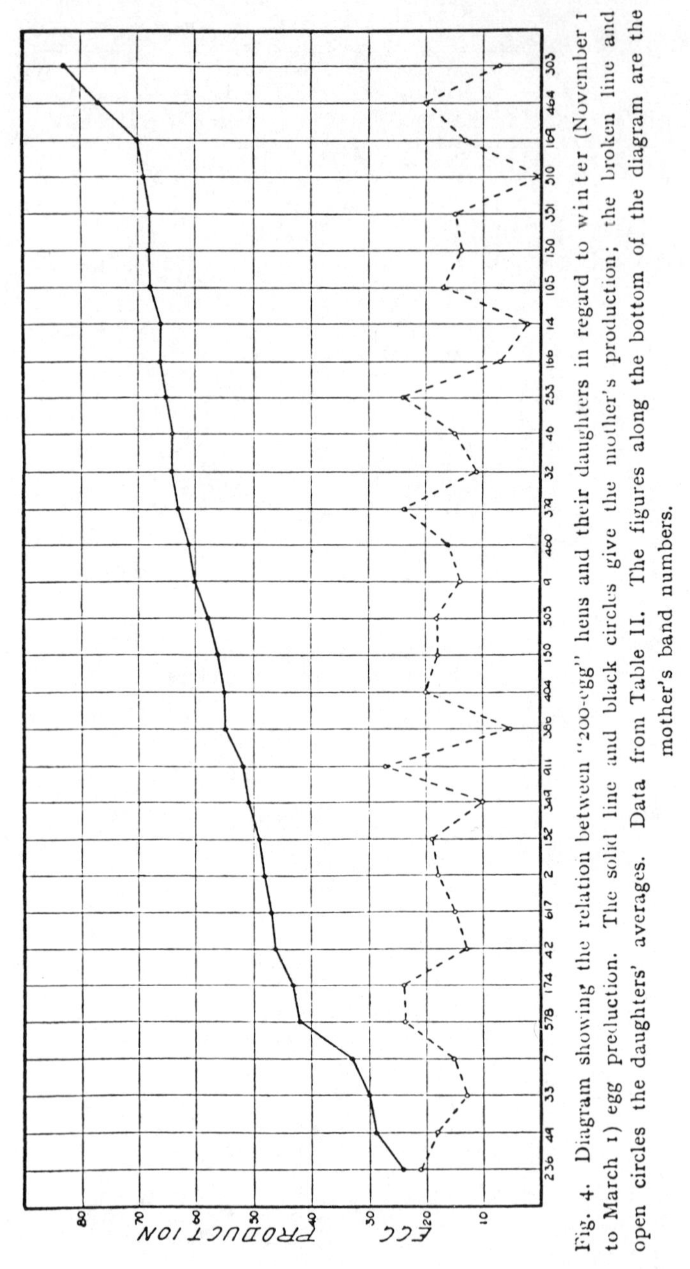

Fig. 4. Diagram showing the relation between "200-egg" hens and their daughters in regard to winter (November 1 to March 1) egg production. The solid line and black circles give the mother's production; the broken line and open circles the daughters' averages. Data from Table II. The figures along the bottom of the diagram are the mother's band numbers.

daughters' average production might be very much lower than their mothers' average production and there still be inheritance of the egg producing ability. This seeming paradox might arise in this way. If unfavorable environmental influences acted in the case of daughters the average production of the whole group of daughters might be considerably below that of the mothers. At the same time the exceptional mother (that is the mother whose production was above the average for mothers) might produce the exceptional daughter (the daughter whose performance was above the average for daughters). Such a condition of affairs would obviously indicate the inheritance of egg producing ability and yet clearly might exist quite independently of the relative magnitude of the averages of the mother and daughter groups as wholes.

To determine whether there is such an inheritance of egg producing ability independent of the group averages it is necessary that the correlation between mothers and daughters in respect to egg production be actually measured. From such measurement it can be told whether on the average the exceptional mother produces the exceptional daughter or whether the exceptional daughter is as likely as not to be the daughter of the mediocre or poor mother.

Such correlations have been determined for each laying period. The net result is to show that in this material there is no sensible correlation between mother and daughter in respect to egg production. In other words there is no evidence that fluctuating variations in this character were inherited in the mass in this experiment. This lack of correlation is shown graphically in Fig. 4 which compares mothers and daughters in respect to winter egg production. If there were an inheritance of fecundity it would be expected that the two zigzag lines in this figure would take a more or less parallel course. This they obviously do not do.

Turning to the question as to whether the daughters of "200-egg" hens were or were not better egg producers than other birds not of such highly selected ancestry (see p. 262 supra) when kept under identical environmental conditions we have the comparative figures given in Table III.

From the table it appears that:

1. With a single exception where the difference is very small (100 bird pens) the mean production of the "unregistered" (less closely selected) birds was in all cases higher than that of the "registered" (more closely selected) birds. In other words the data show that the daughters of "200-egg" hens are certainly not

Table III.

Comparison of the Egg Production of "Registered"[2]) and "Unregistered"[2]) Pullets. November 1—June 1, 1907—1908.

Groups of Birds Compared	Mean or Average Egg Production		Variation in Egg Production			
			Standard Deviation in Egg Production		Coefficient of Variation in Egg Production	
	Nov. 1—Mar. 1	Mar. 1—June 1	Nov. 1—Mar. 1	Mar. 1—June 1	Nov. 1—Mar. 1	Mar. 1—June 1
"Unregistered"—(50 bird pens) . . .	19.38 ±1.18	53.38 ±1.30	17.45±.84	18.16±.92	90.02±6.99	34.02±1.91
"Registered"	15.92[1])±.78	46.83[1])±.90	16.10±.55	18.10±.64	101.14±6.44	38.66±1.54
Difference	+3.46	+6.55	+1.35	+.06	—11.12	—4.64
"Unregistered" (100 bird pens) . . .	21.43 ±1 23	45.65 ±1.36	18.08±.87	19.32±.96	84 37±6.33	42.33±2.45
"Registered"	15.92[1])±.78	46.83[1])±.90	16.10±.55	18.10±.64	101.14±6.44	38.66±1.54
Difference	+5.51	—1.18	+1.98	+1.22	—16.77	+3.67
"Unregistered" (150 bird pens) . . .	17.89 ±.87	47.97 ±1.01	15.78±.62	17.84±.71	88.19±5.51	37.19±1.67
"Registered"	15.92[1])±.78	46.83[1])±.90	16 10±.55	18.10±.64	101.14±6.44	38.66±1.54
Difference	+1.97	+1.14	—.32	—.26	—12.95	—1.47

1) Obtained from the grouped data of Table III and Table IV. The grouping is the cause of the insignificant difference between these averages and those of Table II.

2) For the significance of these terms see p. 261 supra.

higher producers than other pullets of less intensely selected ancestry.

2. The coefficients of variation clearly show that the "unregistered" birds are relatively less variable than the "registered." There is only one exception to this rule and the difference in that case is very small. This means that the daughters of "200-egg" hens instead of conforming more closely to a particular type of egg production, as would on general grounds be expected, actually conform less closely to type than do birds of less closely selected ancestry.

Discussion of Results.

The answer given by the investigations here summarized to the questions stated as the title of this paper is definite and clear. So far as the character fecundity (egg production) in the domestic fowl is concerned long continued and carefully executed experiments give no evidence whatever that there is a cumulative effect of the selection of fluctuating variations.

The facts brought out by this work indicate clearly enough that all birds which have equal records of performance in respect to egg production are not alike in their ability to transmit fecundity. Judged by the performance of their offspring a group of "200-egg" hens, while very homogeneous so far as performance records are concerned, must be very heterogeneous in regard to the constitution of the germ cells relative to the character fecundity. Some "200-egg" hens are apparently capable of producing offspring with very high laying capacity (cf., data in Table II). Other "200-egg" hens exactly similar in all observable respects lack this capability. These facts are of exactly the same order as those which have been brought out by Mendelian work, showing that the constitution of the soma furnishes no certain criterion of the condition or constitution of the germ cells.

But the assumption, tacit or expressed, which lies at the foundation of mass selection methods in practical breeding is that the soma does as a matter of fact give a working criterion of the constitution and potentialities of the germ cells. One breeds from the superior individuals in regard to somatic characters because he expects that the offspring will be superior. Is this expectation well founded? Altogether much evidence is accumulating from widely different sources to show that simple selection of superior individuals as breeders can not alone be depended upon to insure definite or continued improvement in a strain. Some improvement may possibly follow this method of breeding at the very start but the limits both in time and amount are very quickly reached. In support of this view of the possibilities of selective breeding the results set forth in the present paper furnish definite and positive confirmatory evidence. Our experience shows that in order to establish a strain of hens in which high egg production shall be a fixed characteristic it is necessary to do something more than simply breed from high producers. It will of course be understood that our investigations are not stopping at this point. It is proposed to test the conclusions stated in this paper in every possible way.

That greatly exaggerated ideas as to the effectiveness of continued selection of fluctuating variations in improving stock have been widely held during the last half century admits of no doubt. Yet many practical breeders have clearly understood that something more than this was necessary to insure certain and definitely fixed improvement. In this connection a statement from a chapter on "Selection" in an old work[1]) containing much interesting matter concerning current opinion of the time on breeding questions is worth quoting. After outlining briefly the general plan of selective breeding the author goes on to say: "It is not merely by putting the best male to the best female, that the desired qualitites can be obtained; but by other means not clearly defined in the common practice." May it not fairly be said that, as a result of the work of Nilsson, De Vries, Johannsen and others, some at least of these "other means" are now coming to be "clearly defined"?

The results set forth in this paper raise in one's mind some doubt as to the validity of the explanation commonly given for the superiority of present day races of poultry over the wild Gallus bankiva in regard to egg production. It is generally held that the reason for this superiority lies in the continued selection for increased egg production which is assumed to have been practised during the centuries since the domestication of poultry began. Critical examination of this explanation indicates, however, that it has some very weak points. The following considerations are significant in this connection.

[1]) Walker, A. Intermarriage: or the Mode in which, and the causes why, Beauty, Health and Intellect result from certain Unions and Deformity, Disease and Insanity from others: Demonstrated by . . . . . and by an account of corresponding Effects in the Breeding of Animals. American Edition, New York 1839

1. It is an assumption for which we have been able to find little historical warrant to say that selection for egg production was systematically or generally practised with poultry before sometime in the first half of the nineteenth century. Yet there are definite and, by all tests possible to make, authentic records of egg production as high as anything we now know, but made before 1800.

2. The definite experimental results set forth in the present paper do not afford any evidence that it is possible to increase egg production by mass selection methods.

3. There is evidence that the explanation for the superior egg production in the domesticated as compared with the wild Gallus lies in the effect of the environmental influences comprised in the process and conditions of domestication itself. This evidence consists in the known fact that wild birds other than Gallus when put under conditions of domestication have their egg production immediately increased over what it was in the wild state, without the intervention of any selective breeding whatever. We may cite here two instances in support of these statements. The first has to do with the wild Mallard duck (Anas boschas) in captivity. Mr. E. H. Austin[1]), who makes a specialty of domesticating wild water fowl and has had much experience in this direction makes the following statement: "The Mallard duck in the wild state lays 12 to 18 eggs. In captivity it will lay 80 to 100 if the bird in confined at night in a pen and has liberty (in pond or large enclosure) during the day, and the eggs removed daily. As far as my experience goes this is the case with no other variety of duck. The others will desert their nest and stop laying if their eggs or nest are troubled. The Mallard is the original ancestor of the Rouen ducks, and when taken into captivity grows large and coarse in a few generations. It is necessary to constantly use wild drakes to maintain the fine lines and graceful carriage of the wild bird."

A similar condition of affairs has recently been recorded by Duerden[2]) for the ostrich, a bird which has certainly not been selected for egg production during its period of domestication. This investigator states as one of the results of his study of some ostrich egg records that: "The numbers show that ostriches, like poultry,

[1]) Austin, E. H. Original Laying Capacity. Farm-Poultry (Boston) Vol. XIX, p. 347, 1908.

[2]) Duerden, J. E. Experiments with Ostriches. VI — Egg — laying Records of Ostriches. Agricultural Journal (Cape of Good Hope) April, 1908.

will go on laying almost continuously during the breeding season, if the eggs are removed as laid, and the birds are not allowed to sit." He further shows that, just as with poultry, the environmental conditions under which the birds are kept have a marked influence on the egg production in ostriches.

In general we are strongly inclined to the view that the existing evidence indicates that the superior egg production of present-day races of domestic poultry is in the main the result of the action of the favorable environmental influences included in the process and conditions of domestication, rather than an effect of the selection of favorable fluctuating variations through a long period of time.

Summary.

The data discussed in this paper were obtained from two lines of work. The first of these was an experiment in which for a period of nine years hens have been selected for high egg production. No hens were used as breeders whose production in the pullet year had not been 160 or more eggs. The cockerels used were, after the first year of the experiment, invariably the sons of mothers producing 200 or more eggs in their pullet year.

The second source of data was an experiment in which the inheritance of egg production from mother to daughter was directly measured. Records of the pullet year egg production of 250 daughters of hens laying 200 or more eggs in their (the mothers') pullet year were obtained.

Certain of the most important results obtained may be summarily stated as follows: —

1. Selection for high egg production carried on for nine consecutive years did not lead to any increase in the average production of the flocks.

3. There was no decrease in variability in egg production as a result of this selection.

3. The present data give no evidence that there is a sensible correlation between mother and daughter in respect to egg production, or that egg-producing ability (fecundity) is sensibly inherited.

4. In this experiment the daughters of "200-egg" hens did not exhibit, when kept under the same environmental conditions, such as high average egg production as did pullets of the same age which were the daughters of birds whose production was less than 200 eggs per year.

5. The daughters of "200-egg" hens were not less variable in respect to egg production than were similar birds whose mothers were not so closely selected.

23

Reprinted from pages 222–235 of *Int. Cong. Genetics, 4th Proc.*, 1913

THE RESULT OF SELECTING FLUCTUATING VARIATIONS DATA FROM THE ILLINOIS CORN BREEDING EXPERIMENTS[1]

By Dr **Frank M. SURFACE**

There are few biological problems which are attracting more attention at the present time than that regarding the effect of selecting small, fluctuating variations. Until recently the effectiveness of such selection has been accepted almost without question. However, the recent work of Johannsen, Jennings and others has caused many biologists to entertain serious doubts as to the real evolutionary significance of so-called fluctuating variations. If one searches through the literature for clear cut cases in which plants or animals have been modified by the gradual accumulation of small variations he is surprised at the small number which are supported by adequate data. Of these few cases there is one which has been referred to frequently as a classic example, of what can be done by simple selection. I refer to the work of breeding corn (maize) for chemical constitution carried out by the Illinois Agricultural Experiment Station.

Over two years ago while engaged in some corn breeding work at the Maine Experiment Station, the writer had occasion to work out the pedigree tables which are given in the present paper. (cf. pp. 225-229). When displayed in this way, the results of these extensive experiments appeared suggestive of the results which actually come from an attempt to select fluctuating variations. The writer is well aware that no definite conclusions of far reaching importance can be drawn from this data alone. It was for this reason and with the hope of accumulating more definite data that publication has been delayed so long. In view of the interest manifested in this subject at the present time it has seemed worth while to publish these tables together with a brief discussion.

In 1908 the Illinois Experiment Station published[2] the detailed evidence of very careful and long continued experiments in selecting corn with reference to the chemical constitution of the grain. Four definite experiments were carried out simultaneously, viz : (1) Selecting to increase the protein content; (2) Selecting to decrease the protein content; (3) To increase the oil content and (4) to decrease the oil content. For the details of these experiments the reader must be referred to the bulletin mentioned and others by the same Station. For the present discussion it will be sufficient to mention only a few of the more important points regarding the methods used in these experiments.

After some preliminary work there were selected in 1896 one hundred and sixty-three ears from a standard variety of white dent corn. Chemical analyses were made from samples of these ears, showing the protein and oil content of each. From these one hundred and sixty-three ears the twenty-four showing

1. Communication faite à la seconde séance de la Conférence.
2. Ten Generations of Corn Breeding, by Louis H. Smith. *Ill. Agr. Exper. Station Bull.*, No 128.

the highest per cent. of protein were selected for starting the " high-protein " plot. In a like manner the twenty-four ears showing the highest per cent of oil were chosen for the " high-oil " plot. Similarily the twelve ears having the lowest protein content and the twelve ears having the lowest oil content were selected for starting the " low-protein " and the" low-oil " plots respectively. In each case the plots were planted on the ear-to-the-row system. In the following year there were analyzed a number of ears from each of the twenty-four rows, say of the high-protein plot, and from these ears the twenty-four showing the highest per cent of protein, regardless of pedigree, were selected for planting in the " high-protein " plot the following year. Similar methods were employed in the succeeding years and in the other plots. The important point for consideration here is that in each year those ears from a given plot which showed the greatest deviation in the desired direction were chosen to continue that experiment. Thus it was an experiment in selecting fluctuating variations.

The success of the experiment in accomplishing the desired results are most marked. Thus starting with an average protein content, in the original one hundred and sixty-three ears, of 10.92 per cent they were able in ten years to increase the average per cent of protein to 14.26. On the other hand in the low protein plots the percentage was reduced to 8.64. Similar results were accomplished with the high and low oil plots. The yearly fluctuations in the different plots are shown in tables 1 and 2 which are taken from tables 5 and 6 of the Illinois bulletin No 128.

TABLE 1.

Average per cent. of Protein in the Crops harvested.

YEAR.	HIGH PROTEIN PLOT.	LOW PROTEIN PLOT.
1896	10,92	10,92
1897	11,10	10,55
1898	11,05	10,55
1899	11,46	9,86
1900	12,32	9,34
1901	14,12	10,04
1902	12,34	8,22
1903	13,04	8,62
1904	15,03	9,27
1905	14,72	8,57
1906	14,26	8,64

TABLE 2.

Average Per cent. of Oil in the Crops harvested.

YEAR.	HIGH-OIL PLOT.	LOW-OIL PLOT.
1896	4,70	4,70
1897	4,73	4,06
1898	5,15	3,99
1899	5,64	3,82
1900	6,12	3,57
1901	6,09	3,43
1902	6,41	3,02
1903	6,50	2,97
1904	6,97	2,89
1905	7,29	2,58
1906	7,37	2,66

That the striking results shown in these tables were not due to the effect of environmental circumstances is clearly shown by the analyses from the so-called " mixed protein " and " mixed-oil " plots. In these plots kernels from the high and the low strains were planted in the same hills. Subsequent analyses showed that under these conditions the various strains maintained their respective chemical characteristics. Thus there cannot be the least doubt but that certain characters were fixed[1] in these various strains by the selection practiced.

At the time that the selections were made a careful record of the pedigree of each ear was kept. These pedigrees are, of course, for the maternal side only, since hand pollination was not practiced. In the appendix to bulletin No. 128 all the analyses for the ten years are given and arranged in such a way that it is possible to trace out the pedigree of each individual ear. It is from this data that the following pedigree tables (Tables, 3, 4, 5, and 6) have been constructed. In the following discussion it will be advantageous to take each plot separately. This may be done in the order in which they occur in the bulletin.

High Protein Corn.

As stated above twenty-four ears containing the highest per cent of protein were selected from the one hundred and sixty-three ears analysed in 1896. These were given registry numbers from 101 to 124 inclusive as shown in column one of table 3[2].

The next season four sound ears were analysed from each of the twenty-four rows. From these ninety-six ears, the twenty-four again having the highest per cent of protein were selected for planting. The distribution of these selected ears among the twenty-four original ears is shown in column two of table 3. For example it is seen that ear No 124 produced two ears (nos 216 and 209) which were among the first twenty-four as regards protein content. Ear No. 123 on the other hand failed to produce any ear (so far as the ears analysed showed) sufficently rich in protein to be included among the first twenty-four. In this way it is seen at once that eight of the twenty-four original ears fail to be represented in the second generation while eight other ears contribute two ears each for planting in the following year. Exactly the same selection was practiced in the second year and the resulting selected ears are shown in the third column of table 3. Of the sixteen original ears represented in the second generation, only one, viz : No. 116, was dropped out in the third generation. In the next generation there is a very significant dropping out of some of the original lines. Thus in this fourth generation, only 9 of the original twenty-four ears are represented by progeny. Five of the original lines contribute twenty of the twenty-four ears, or 80 per cent in this generation while two lines, viz : 106 and 112 contribute fourteen ears or nearly sixty per cent. Thus at the end of the fourth generation it is clear that certain of the original lines have a much greater tendency to produce ears with a high per cent of protein. By simply selecting on the basis of the protein content of the individual ear, for four years, 70 per cent of the original lines have been dropped.

1. This point is also brought out in a recent paper by L. H. Smith, " Increasing Protein and Fat in Corn ", *American Breeders, Assoc. Report.*, vol. VI, pp. 5-11, 1911.

2. For convenience these ears will be called the first generation of high-protein corn.

TABLE 3.

Pedigree Chart of « High-protein » Corn.

GENERATION N°.										
1	2	3	4	5	6	7	8	9	10	11
101										
102 —	215 —	320 —	410 —	502						
103 —	208 —	314. .	424							
			409							
104 —	214. .	316 —	421							
		310								
105										
			401 —	514						
		306. .								
106. .	222		405							
	213. .	315 —	418							
			417							
		319. .	414							
			416							
107. .	219 —	301								
	223									
			415							
	206 —	321. .								
108. .			406							
	217									
109										
			407 —	510						
110			420							
		311. .								
			411. .	506 —	604					
111				513						
		312								
	212. .									
		313								
			404							
					614					
112. .		309. .	402 —	515. .	606					
					603					
			408							
			422							
	205 —	317. .	412. .	503						
				509 —	613 —	705 —	822			
113. .	210									
	220 —	324								
114 —	204 —	303								
115 —	224 —	304								
116 —	202									
117										
118 —	218 —	302								
119. .	221 —	307 —	403 —	511						
	201 —	305								
120 —	203 —	308								

TABLE 3 (*Suite*).

Pedigree Chart of « High-protein » Corn.

Generation No.										
1	2	3	4	5	6	7	8	9	10	11
							808.	901		
								908		
						706. .	813			
					609. .		801			
				512. .		702			1007	
									1002	
								912. .	1014	
					608				1019	
					612 —	711	814. .			
								925		
									1001	1110
									1020 .	1103
					601 —	710. .	810	920. .	1008	1115
										1122
			423				821		1013	
				505				916		
								902		
121. .	207 —	323. .	415. .				811. .	905	1009	1102
	211					719		914. .	1016	1114
				501		721			1004 .	1107
					602. .	715	819		1021	1119
						712				
								906		
							806. .		1017	1113
								924. .	1024	1101
						713. .			1012 .	1120
									1005	1108
							802			
				507. .				903		
						704	820. .	917		
								910		
				508	607. .					
						714				1111
									1010 .	1106
										1123
						716 —	807			1118
							818. .	911		
								922. .	1003	
						717. .				
									1015	1105
							804	915	1022 .	1112
					605. .	707	812. .	919		1124
						720		923		1117
							805			
						708	815			
						709. .		907		
					611. .		803. .	913	1011	
							817	904		1104
						703	809. .	918. .	1018	1109
						701		921	1006 .	1116
									1025	1121
122										
123										
124.	216 —	318								
	209 —	322 —	419 —	504 —	610. .	718 —	816 —	909		
						722				

TABLE 4.

Pedigree of « Low protein » Corn.

GENERATION N°.										
1	**2**	**3**	**4**	**5**	**6**	**7**	**8**	**9**	**10**	**11**
102										
103	208									
	203									
			407	501						
104	201	309	405	509						
			410	512						
				504						
					606	706				
						701		918		
		313	402	514	605		815	925		
						702	805	904		
105	207	316				713	816			
		302	416		603	720	819			
			415			719				
						705				
	212									1107
										1119
									1006	
106	211	311								1102
		303		505	612	708	812	917	1011	1114
	210					714	820	912	1023	
		314	403							1112
								923		1105
					602	715	804		1018	1124
									1004	1117
									1009	
						712	811	906		1110
							817			1103
				506					1016	1122
							818	921	1021	1115
								910		
							808			
									1005	
								919	1024	
							821	924		1106
					613	704			1017	
								915		1111
									1012	1118
							806			
										1123
							810	905		
								922		
									1007	1108
									1002	
							802	911		1120
								902	1014	1101
									1019	
										1113
		308	401	511			801			
	205		406	503	601	707	813			
		304	413		611					
107	202				614	722				
	209	307	404	502		709				
	204	312		507	609					
					610	703	822			
					604	717				
						710				
		305	411	510	608	718		903		
						716	807	909		
						721	814	916		
									1001	
						711		908		
							803	914	1013	
									1008	
								901		
									1020	1104
								907		1116
									1010	
							809	913	1003	1109
								920		1121
108		315	412	513	607				1022	
		301							1015	
109	206	310	414							
110			409							
111		306	408	508						
112										

TABLE 5.

Pedigree table of « High-Oil » plot.

GÉNÉRATION N°.

1	2	3	4	5	6	7	8	9	10	11
101										
102								909		
103							810. .	914		
104 —	209							925		
105 —	219							917		
106 —	203						812. .	922		
107 —	224 —	308 —	403 —	506		709. .		907		1107
108										1102
109 —	211 —	302							1008 .	1114
110 —	221							921	1001	1119
					601		814. .	916. .	1013	1103
				511. .	605			903		1110
	210 —	303. .	407. .		609. .		821		1020 .	1115
			409	508. .	608	715. .	817	911	1004	1122
							808. .	904. .	1009	
111. .	213					707			1021	
					603				1016	
						718. .	801			
	216						805			
	214 —	311. .	408			717 —	802. .	901		
			404 —	507. .		714		912		
										1104
									1003	1109
								908		
									1010 .	1116
								918. .		
						702 —	816. .	915	1015	1121
						708			1022	
					610. .		803			
							806			
							809			
112 —	217									
						713. .	811	919		
							813. .	906		
							815	924		
							819			
113. .	201			513-						
		306 —	401. .	505						
	212.									
		307 —	412 —	503 —	606					
			402 —	502	613 —	701				
114. .	215 —	305. .	411. .	514						
				509. .						
	206				614. .	706				
						719			1002	
								905		1106
					611. .	721 —	822. .	902. .	1007	1111
						712			1014 .	1118
									1019	1123
										1105
									1017	
										1112
	204 —	310							1024 .	
								920. .		1117
115. .	202 —	312					804. .	913	1012	1124
	208 —	304 —	406 —	501					1005	
						703				
					602. .	710. .	807			
116 —	223					720		923		
							820		1011	
117 —	220						818. .		1006	1101
				512. .	612. .	711		910. .	1018 .	1108
						716			1023	1113
						722				1120
118. .	207 —	309 —	410. .							
	222				604. .	705				
						704				
				504 —	607					
119										
120 —	218 —	301 —	405 —	510						
121										
122 —	205									
123										
124										

TABLE 6.

Pedigree Chart of « Low-Oil » Corn.

GENERATION N°.										
1	2	3	4	5	6	7	8	9	10	11
101										
102 —	202									
103										
104									1004	
			405							
	201 —	312. .	412					908. .	1009	1101
105. .					602 —	718 —	806. .	901	1016 .	1108
	209 —	303 —	402					915	1021	1113
				507. .						1120
						716. .	817			
					612. .	705	814 —	909		
						706				1102
		314 —	408. .				813	911	1002 .	1107
						710	804. .			1114
								906, .	1007	1119
				510						
							819			
	207. .	316			606. .				1019	1106
									1014 .	1111
				512. .		722. .				1118
								902	1001	1123
		311 —	406. .				811. .	917	1008	
					611 —	709		914.	1013	1105
106. .					613					1112
	212			502. .			809		1020 .	
					614 —	702				1117
								913	1005	
								924		1124
							815. .		1012	
								920. .	1017	
	203 —	305								
		301 —	404 —	513					1024	
	205. .	309 —	416. .	509 —	601					
		308 —	403	511 —	604. .	703				
						704				
						714				
						708 —	816. .	905		
								922		
107										
		315 —	415	508 —	609. .	712	822	903		
							818. .	916		
	204. .					720. .	805	923		
							802			
			409. .				810. .	912	1010	
								921		
108. .		307. .				711	821		1003	
								918. .		
				506 —	607. .	715	812. .		1015	
				505		719. .		907	1022	
			410. .				808	925		
	210			503 —	603 —	707	803			
109 —	208. .	313 —	413 —	514 —	608					
		302 —	401 —	501						
										1104
									1006 .	1109
						701				1116
					605. .	713. .	801	904	1011	1121
							807. .	910. .		
			407 —	504. .		717				
	211	310. .					820	919	1023	1103
110. .			411		610	721				
	206. .	304								1110
									1018 .	
		306 —	414							1115
										1122
111										
112										

In the next or fifth generation three more of these original lines were dropped[1]. Here again one finds that two of the original lines, viz : 112 and 121, contribute eleven of the fifteen selected ears or over 75 per cent. Line No. 106 which had six ears in the fourth generation failed to produce more than one ear good enough to come inside the first fifteen.

Passing to the sixth generation one finds that three more of the original lines are dropped from the contest. There are only three of the original twenty-four ears represented by progeny in this, the sixth generation, and one of these (No. 124) contributed only a single ear.

In the seventh generation these same three lines are represented. However, the superiority of line No. 121 to produce ears high in protein is clearly evident. Nineteen of the twenty-two ears, or over 86 per cent, come from this one line.

In the eighth generation there is practically no change in the relationship of these lines. Twenty of the twenty-two ears are furnished by line No. 121. In the ninth generation line No. 112 drops out and line No. 124 fails to secure a place in the tenth generation. *It thus happens that in the tenth and eleventh generations all of the high protein corn is the offspring of a single ear* viz : *No.* 121.

It is to be remembered that this condition has been brought about simply by selecting each year those individual ears which showed the highest per cent. of protein. Everyone who has had any experience with selection of this kind knows that many of the original lines are always dropped out as the work proceeds. But it is rather surprising to find so striking an example of the superiority of a single line[2].

Before considering the other experiments it will be well to examine the pedigree of this line a little more closely. In the first four generations there is nothing remarkable in its history. Up to that time there are several other lines which give more promise of producing high protein ears than this one.

Line No. 112 has eight ears in the fourth generation and line No. 106 has six ears, while No. 121 has but two. One of these ears (No. 415) however, had within it the ability to produce ears high in protein content. In the fifth generation this ear produced five of the fifteen ears planted. This is a larger number of ears high enough in protein to be selected, than had been produced by any one ear up to this time. So far as can be learned from the published records there are no special cultural reasons why the progeny of this ear should have been higher in protein than the progeny of other ears.

In the sixth generation of line 121, ear No. 512 produced two ears and ear No. 507 produced six ears out of a total of fourteen planted. It should be stated here that in this year (1901) a severe drought injured the corn very much and had considerable influence on the resulting selection. The following quotation from the Illinois Bulletin No. 128 will make this plain : (page 469).

" In the case of the high-protein plot the damaging effect of this drought was so pronounced as to render the crop almost a total failure. The yield of ear corn amounted to only about six bushels per acre and consisted mostly of

1. It should be mentioned here that for the fifth generation only fifteen ears were selected and for the sixth generation only fourteen ears. For the years following the number of ears selected ranged from twenty-two to twenty-five.

2 In the tenth and eleventh generations the plan of detasseling alternate rows and saving seed from these only was put into effect. Also in these years a much larger number of ears were analyzed from each mother row. This change in method could not have effected the results brought out here, because, line No. 121 had demonstrated its superiority several generations before this change was made.

mere nubbins. On account of the scarcity of ears, it was impossible to follow the regular system of sampling, so the entire product from each plot-row was collected and all of the sound ears and even many nubbins were selected for analyses in order to obtain the results of the year and the sort of seed with which to maintain the experiment. The low protein plot did not suffer so badly from the drought, so that here the sampling and selection were made as usual

Ears No. 507 and 515 were planted in the " Special High-Protein Plot " and all the ears selected from these two lines were from this plot. In the appendix of the bulletin mentioned, the analyses of twenty-seven ears from No. 507 (line 121) are given. Of these, six were selected. Similarily from ear No. 515 (line 112), the analyse of forty-five ears are given but only three were selected. None of these three ears were able to get progeny in the next generation while in striking contrast *every one* of the six ears from No. 507 produced ears good enough to be selected the following year. Further these latter ears produced practically *all* of the High-protein corn from this time on.

Certainly ears No. 507 and 515 at least had the same opportunity to produce high protein ears. The results as shown above however are strikingly different.

Looking at the pedigree of line 121 alone it may be said to fall into two parts. The first of these, covering the first four generations is characterized by mediocrity in protein productions. The second part i. e., after the fourth generation, is characterized by the production of a large number of ears with high protein content.

Evidently something happened to ear No. 415, or perhaps to ear No. 507 which produced a prepotency towards high protein production. What this something may have been, can only be conjectured. It may have been of the nature of a mutation. This is perhaps the first suggestion that occurs to one. In this case there are certain possible contributory factors which may have been operative. The chief of these are to be found in either selective fertilization or in the proper amount of inbreeding, At this time the larger amount of pollen in the field was coming from two or three lines. These, of course, had been crossed with other lines but nevertheless they were pure bred on one side. We know too little of the effect of such inbreeding to make more than a suggestion.

The suggestion is certainly close at hand that in the third or fourth generation of line No. 121 a particularly happy combination or germinal plasma occurred. As a result of this combination a line which was previously only mediocre in its protein producing ability suddenly acquired a marked increase in this direction. This ability to produce a large number of high protein ears has evidently been transmitted to later generations.

The phenomena of propotency which it seems to me is displayed in this instance is one which has been discredited largely by scientific breeders in recent years. Pearl[1] however, has recently shown that in poultry the offspring of certain matings transmit for several generations at least, the ability to produce either a high or low degree of egg production as the case may be.

One more point must be considered here. From table I (page 225) it is seen that starting with a protein content of 10.92 per cent. at the end of the

1. Inheritance of Fecundity in the « Domestic Fowl », *Amer. Nat.*, vol. XLV, June 1911.

third year the protein content was only 11.46 per cent. or a gain of. 54 per cent. in four years. However, the next year the protein content jumped to 12.32 per cent., or a gain of. 86 per cent. in one year. Now referring back to table 3 it is seen that it is just at the end of the third years selection that there is a great reduction in the number of lines represented. Thus only six of the original twenty four lines are represented in the fifth generation. Also it is just here that line No. 121 begins to show its prepotency. The next year (1901) there is a very much larger increase in the average per cent. of protein, viz : 1.80 per cent. for the year. It has already been stated that climatic conditions caused an exceptionally high per cent of protein in this year. The next year (1902) the protein content fell back again and then rose with some irregularities.

The most interesting feature here, however, is that so long as a comparatively large number of lines are represented in the pedigree the average protein content remains almost stationary. When the number of the lines begins to be diminished the average protein content begins to be increased much more rapidly. This may of course, be a coincidence but it is a point which needs more analytical study.

Before discussing this question further it will be well to examine very briefly the remaining three experiments. In no one of these are the results so striking as in the high protein plots. However in a general way they point to the same conclusions.

Low Protein Plots.

Table 4 shows the various lines of the low protein plots arranged in the same way as the high protein pedigrees. It will not be necessary to enter into the details of these tables. They present many interesting things to anyone who cares to study them. Only a few of the more important conclusions are pointed out here.

In the first place, it is seen that starting with twelve original lines there are only two of these represented in the last generation. Only three lines persist beyond the sixth generation. Of the two lines which have progeny in the last generation, one is much superior to the other in its ability, to produce ears low in protein. This has not been true of its whole history. In the first seven years, line No. 106 produced only a moderate number of ears low enough to be selected. After the seventh generation. however, this faculty appears to have increased very greatly.

Comparing this table with column 3 of table 1, it is seen that it was not until 1902, or the seventh generation, that any very great or permanent decrease in the per cent. of protein was brought about. Here again the most marked improvement in the direction sought occurs only after the number of lines has been very materially reduced.

The results of the low protein experiment while not quite so striking as those of high protein plot are nevertheless along the same general lines and tend to confirm the conclusions drawn from it.

High Oil Plots.

Table 5 gives the pedigrees for the high oil plots. From this table one may briefly note the following points :

1. Of the twenty-four ears used to start this experiment in 1896 only eight are represented by progeny in the third generation;

2. At the end of the sixth generation all but three of the original lines have been dropped;

3. These three lines are maintained to the end of the experiment. The number of selected ears in the later generation are not, however, equally divided between these three lines. Line No. 111 is by far the most prolific producer of ears with high oil content;

4. Again certain apparently crucial points in the history of these lines are noted. Thus in Line No. 111, ear No. 609 produced an exceptionally large number of ears with a high oil content. Similarily in line No. 118, ear No. 710 produced four good ears while seven other ears from the same line in the same generation failed to produce any which were good enough to be selected.

Low Oil Plots.

As shown in table 6, twelve ears were selected for starting the experiment in decreasing the oil content. From this table we note:

1. That only three of the original lines are represented in the seventh generation;

2. Of these three only two, viz: 106 and 110, are represented in the eleventh generation. Line No. 108 is maintained up to the eleventh generation and then dropped;

3. Of these two lines, No. 106 is by far the better, and contributed sixteen of the twenty-four ears to the last generation;

4. Again it is noted that certain ears have a marked tendency to produce a large number of ears with a low oil content. Examples of this are seen in the seventh generation in ears No. 722, 720, and 719.

Discussions and Conclusion.

Taking into consideration the results of all four experiments as displayed here it is clearly seen that one of the most striking effects of the selection practiced has been to reduce the number of lines. Two of the experiments (Tables 3 and 5) were started with twenty-four ears each. At the end of five years the high-protein corn shows progeny from only three of the original ears, and the high-oil from four of the original ears. At the end of ten years the high-protein corn showed progeny from only one of the original ears and the high-oil from three of the original ears. So far the results are in accord with the genotype conception of Johannsen as applied to non-self-fertilized organisms. However, if nothing but the isolation of favorable genotypes had taken place the extremes of protein production would not have been greatly changed. This, however, did take place here. In the later years we find many individual ears with a per cent. of protein far beyond anything which occurred in the earlier years. In this respect these results parallel the classical case of de Vries's selection of buttercups. In this latter case the extreme was moved far beyond what it was before the selection. Unfortunately in the case of de Vries's work pedigree data is not available.

Table 7 shows for each year the maximum and minimum per cent. of protein in the ears which were selected for planting the high-protein plots.

TABLE 7.

Table showing the maximum and minimum per cent. of protein in the ears selected for planting in the high-protein plots.

YEAR.	MAXIMUM.	MINIMUM.
1896	13,87	11,89
1897	13,62	11,89
1898	14,92	12,35
1899	14,78	13,19
1900	15,71	14,01
1901	16,12	14,93
1902	15,01	13.68
1903	17,33	14,60
1904	17,79	15,83
1905	17,39	15,52
1906	17,67	15,16

From this table it is clear that the maximum has been moved permanently in the direction of the selection. To obtain a more reliable basis for comparison we may average the maximum per cent. of protein in the first three years, and in the last three years. These averages are 14.14 per cent, and 17.62 per cent., respectively, or a difference of 3.48 per cent. Further it is seen that in the last four years the minimum per cent of protein in the selected ears is greater than the maximum during the first two years.

Clearly we are dealing here not merely with the isolation of a genotype but with a definite evolutionary change. The whole variation polygon has been moved definitely in the direction of the selection[1].

Yet I think that no one can study table 3 and still maintain that it was the simple selection of fluctuating variations that brought about the change in protein production. Certainly something was acting in line No. 121 which was not affecting the other lines.

The recent interesting experiments of Shull and East in dealing with homozygous strains of corn may have some bearing in this connection. These authors have shown[2] that when two strains of corn which have been inbred for several generations are then crossed the resulting progeny are far superior to either parent in yield. It is quite possible that a somewhat similar explanation would account for the great development of line No. 121, after the fourth generation.

It is of interest to note here the similarity between these experiments and the selection work with poultry at the Maine Agricultural Experiment Station. It has been shown[3] that after nine years of intensive selection of poultry for increased egg production the average production of the flock was not changed. A possible explanation of the different result reached by these two similar experiments may be in the fact that the corn plot was started with only twenty-four individual ears, while in the poultry experiments, seventy birds were used in the first year's breeding. In the succeeding years a very much larger number of

1. This question as to the amount of shifting of the entire variation polygon was studied to some extent, but it was deemed that the published data was not sufficient to draw trustworthy conclusions.
2. Shull (G.-H.). A pure line method in corn breeding, *Rpt. Amer. Breeders Assc.*, vol. V, pp. 51-59 — East (E.-M.). *Amer Nat.* Vol. XLIII p. 173-182, 1909.
3. Pearl (R.) and Surface. F. M., U.S. Dept. Agr., Bur. Anim. Ind. Bul. 110. pt. 1, 1909, *Zeit. f. Induct. Abst.-u.-Vererb-Lehre.*, Bd. 2, 1909, pp. 257-275.

birds were used, while in the corn experiments there were never more than twenty-five ears planted in one plot. It would have been a much more difficult task to have isolated the high genotypes if such existed among the poultry.

In conclusion it should be said that too much credit cannot be given to the Illinois Experiment Station for the careful manner in which these experiments have been carried out. The experiments were begun with the practical object of finding out whether the chemical composition of corn could be modified by selection. The results showed unmistakeably that it could. Whether the theoretical conclusions were correct or not, the fact remains that the methods used were such as to obtain the desired results.

Part VI

THE GRAND SYNTHESIS: THE SYNTHETIC THEORY OF EVOLUTION

Editor's Comments on Paper 24

24 **WRIGHT**
The Relation of Livestock Breeding to Theories of Evolution

Mathematical models that described the effect of selection, mating systems, mutation, genetic drift, and migration on the frequency and distribution of genes in populations were developed between 1918 and 1932 by Sergei Chetverikov (1926, 1927), R. A. Fisher (1918, 1930, J. B. S. Haldane (1924–1932, 1932), and Sewall Wright (1921, 1931, 1932). These population genetics models enabled biologists to quantitatively compare the relative importance of selection, mutation, and genetic drift in bringing about genetic change in populations. The mathematical modeling of evolutionary processes in populations by Fisher, Haldane, and Wright (1) helped convince scientists that very small selection pressures were capable of genetically changing populations rapidly enough to have caused the changes observed in the fossil record and (2) eliminated many competing theories by showing that Lamarckian, orthogenetic, and other hypothesized factors were not logically essential for evolution to occur (Provine, 1978).

Many of the classic papers on the evolutionary dimensions of population genetics have been published in the Benchmark volumes *Stochastic Models in Population Genetics* (Li, 1977) and *Evolutionary Genetics* (Jameson, 1977*b*), and more will be published in a forthcoming volume on issues in population genetics. The role that population genetics theory has played in the development of selection theory will also be explored in the forthcoming *Benchmark Papers in Systematic and Evolutionary Biology* volumes on natural selection and sexual selection.

The synthetic theory of evolution that developed during the 1930s, 1940s, and 1950s provided an explanation of phylogeny, ongoing evolution, and speciation that incorporated scientific knowledge concerning wild populations gathered by naturalists, and the experimental manipulation of populations of laboratory, domesticated and wild organisms, and that was based on the population genetics models of Mendelian populations constructed by Fisher, Haldane, and Wright (Provine, 1978). The development of the synthetic theory of evolution involved many scientists including Dobzhansky (1937), Huxley (1942), Mayr (1942), Simpson (1944, 1953), and Stebbins (1950).

The synthetic theory of evolution emphasizes that there is a tremendous amount of genetic variability present in natural populations that selection can operate on, that gene interaction and coadaptation are important dimensions of the evolutionary process, and that genetic recombination (sexual reproduction) rather than mutation is the immediate cause of virtually all the genetic variation that appears within a population. The almost universal success of artificial selection experiments supports this last conclusion (Wright, 1977).

The historical development of Sewall Wright's theories concerning evolutionary processes demonstrate the important role that artificial selection and breeding experiments have played in the development of modern evolutionary theory. Sewall Wright worked as a geneticist conducting animal breeding and selection experiments in W. E. Castle's laboratory from 1912–1915 and as senior animal breeder (1915–1925) working in the Animal Husbandry Division of the U.S. Department of Agriculture. These animal-breeding experiences were in large part responsible for the emphasis Wright has placed on modeling the effect that population size, population structure, and the interaction of genes have on genetic evolution (Paper 23; Provine, 1971; Mayr, 1973). Wright has called his theory of evolution, which incorporates random drift (genetic drift or sampling error), intrademe and interdeme selection, the "three-phase shifting balance" theory. Sewall Wright has recently summarized his view of population genetics and evolutionary theory (Wright, 1968, 1969, 1977, 1978). It is appropriate that Sewall Wright's personal account of the relation of livestock breeding to the history of theories of evolution (Paper 24) be the final paper reprinted in this volume.

The role that artificial selection experiments have played in the development of evolutionary theory since 1918 will be explored in forthcoming Benchmark volumes on quantitative genetics, plant breeding, and animal breeding. Some of the classic selective breeding experiments that have provided insight into the evolutionary processes involved in the domestication of plants and animals are reprinted in the Benchmark volumes *Genetics of Speciation* (Jameson, 1977a) and *Hybridization* (Levin, 1979). Additional classic papers will be published in the forthcoming *Benchmark Papers in Systematic and Evolutionary Biology* volume on the domestication of plants and animals by selection. Some of the classic papers attempting to apply artificial selection theory to human population in order to socially direct future human genetic evolution have been reprinted in *Eugenics Then and Now* (Bajema, 1976).

Artificial selection experiments continue to provide insights into evolutionary processes (see, for example, Pollak, Kempthorne, and Bailey, 1977; Wright, 1977). Knowledge about evolutionary processes has been and is still being used by scientists who are attempting to bring about an increase in food production by the selective breeding of domesticated plants and animals (see, for example, Lush, 1945; Frankel, 1974; Frankel and Hawkes, 1975; Pollak, Kempthorne, and Bailey, 1977).

24

Reprinted from *J. Anim. Sci.* **46**:1192–1200 (1978)

THE RELATION OF LIVESTOCK BREEDING TO THEORIES OF EVOLUTION[1]

Sewall Wright

University of Wisconsin[2], Madison 53706

INTRODUCTION

The debt owed by quantitative geneticists to today's guest speaker is immeasurable. We quantitative geneticists are much concerned with the covariances among relatives, so I thought this introduction should include information on near relatives as well as individual performance data. For pedigree evaluation purposes, I thought it necessary to tell you a bit about Dr. Wright's father. Philip Green Wright was born in 1861, and took a Master's Degree at Harvard in Economics. Most of his academic career was spent at Lombard College in Galesburg, IL. There he taught Economics, Mathematics, Astronomy and English, and was Director of the Gymnasium. He was a minor poet and would probably have to be judged unsuccessful in this endeavor, except that one of his student's, Carl Sandburg, enjoyed a bit of success. Another important relationship is that of full-sibs. Dr. Wright's two brothers were successful and able men. Quincy was a Professor of Political Science at the University of Chicago. Theodore was Director of the Aircraft Resources Office, 1942 to 1944, and had primary responsibility of overcoming the lead of the Axis powers in military airplanes. Later he became Vice-President-in-Charge-of-Research at Cornell University. Positive assortative mating occurs in humans, thus information on a spouse is of interest. Mrs. Wright was also a geneticist with a Master's Degree in Zoology. She taught at Smith College for 3 years, during which time she started the first course in Genetics there. For progeny test information, I offer that the Wrights had three children – two sons and a daughter – all of whom took advanced degrees.

Finally, we must consider individual performance. Professor Wright took a B.S. degree at Lombard College in 1911, having successfully passed the course in Calculus and Analytical Geometry given by his father. His M.S. degree was taken at the University of Illinois in 1912 where he studied the microscopic anatomy of a flatworm. Upon completion of his Master's Degree, he studied genetics under Dr. William E. Castle and received his Doctor of Science in 1915. The next 10 years were spent as Senior Animal Husbandman with the U.S. Department of Agriculture in Washington. From there he went to the University of Chicago, and was a member of that faculty from 1926 to 1954. In 1955, he accepted an appointment as the L. J. Cole Professor of Genetics here at Wisconsin, and, since 1960 has been an Emeritus Professor of Genetics. Dr. Wright has received honorary degrees from a number of prestigious universities and many awards, the foremost of which is the National Medal of Science.

His classical series of papers on Systems of Mating, published in 1921, provides the theoretical foundation for plant and animal breeding, our main interest here. But we should also be cognizant of his contributions in other areas of knowledge. His method of analysis of correlated variables, quite familiar to us here, is now widely used in the social sciences. His work on pigment genetics led him to a theory of genes as producers of enzymes, thus anticipating in many ways the work of Beadle and Tatum which initiated modern biochemical genetics. Every student of evolution theory is familiar with his recognition that stochastic changes and the effect of subdivision of populations can be important evolutionary determinants. He has emphasized many times that the evolutionary path of a population is determined by a balance among the various simultaneously-acting forces that change gene frequency. Dr. Wright's first scientific contribution was in 1912. Now some 65 years later, we quantitative geneticists are eagerly awaiting Volume 4 of his encyclopedic treatise – *Evolution and the Genetics of*

[1] Invited paper presented at the 69th Annual Meeting of the American Society of Animal Science, Madison, WI, July 25, 1977.

[2] Department of Genetics.

Populations.

Dr. Wright was a member of the American Society of Animal Production for some 25 years and made several contributions to the annual Proceedings. Dr. Wright, although the Society may have changed its name since you were last a member, we sincerely want to welcome you home.

THEORIES OF EVOLUTION

It is appropriate to begin with Darwin's theory of evolution by natural selection (1859). Direct evidence from nature was not available. The only direct evidence of its possibility was from the results that had been obtained from artificial selection by British livestock breeders during the preceding century. Darwin discussed this at considerable length.

He also discussed whether advances had occurred typically by abrupt steps (as in polled breeds of cattle and in the short-legged Ancon sheep) or had been built up gradually by selection of slight quantitative variants. He decided on the latter—"without variability nothing can be effected: Slight individual differences, however, suffice and are probably the chief or sole means in the production of species."

The opposite alternative was adopted in extreme form by deVries near the beginning of this century (1901–1903). He held that species arise from single mutations that are responsible both for the character changes and for the reproductive isolation that define a new species.

The mutations on which he largely based his theory, those that he observed in the American plant, Oenothera lamarckiana, that had escaped from cultivation in the Netherlands, turned out later to be chromosome aberrations, mostly the presence of an extra chromosome. This proved to be transmissible only by the ovules, so that the mutants existed only as segregants without the reproductive isolation of a true species.

It came to be recognized that certain balanced chromosome changes (tetraploids and translocations) play an important role in the origin of species by leading to the essential reproductive isolation, but are of little significance in character change. The unbalanced ones cause drastic character change, but are in general unfixable and soon lost. I will be concerned here with the transformation of characters within species, not with the origin of species.

After the rediscovery of Mendelian heredity in 1900, most geneticists, led by Bateson (1909) in England and Morgan (1932) in America, thought of species as fixed with respect to "wild type" alleles at most loci. They held that evolution consisted in the replacement of such alleles by very rare mutations that happened to be favorable.

The course of fixation was worked out by Castle in 1903 in a special case. The process was worked out systematically by J. B. S. Haldane under a great variety of genetic conditions in a series of papers from 1924 summarized in his book, "The Causes of Evolution," in 1932. He also dealt with quantitative variation and such complications as gene interaction, but emphasized rare major mutations, probably from considerations of available time. He recognized the principle, painfully familiar to livestock breeders, that selection for any one character restricts the possibility of selection for any other because of limited reproductive excess. He estimated later (1957) that the process of fixation of any gene, major or minor, required some 300 generations. Even geologic time was not enough to fix a very large number of genes with very slight effects. Figure 1 illustrates the course of fixation of a recessive mutation under two conditions with respect to heritability.

At the time of the rediscovery of Mendelian heredity, Darwinian evolution from quantitative variability was being advocated in England by the biometricians under the leadership of Karl Pearson. They strenuously rejected Mendelian heredity in favor of Galton's law of

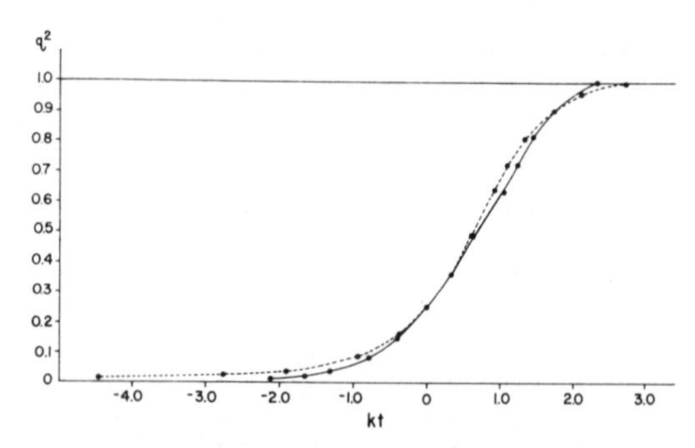

Figure 1. Comparison of the courses of change of the frequencies of recessives (q^2) under directional selection, where variability is entirely genetic (solid line) or is so nearly entirely nongenetic that the selection intensity is treated as constant (broken line). For ease of comparison, the rates are taken as the same as $q = .5$ ($q^2 = .25$). From Wright (1977, figure 6.2).

ancestral heredity (1889) and thus rejected Bateson's (1909) ideas on evolution. The acrimonious debate that ensued probably delayed experimental population genetics for several decades in England, that is, until Mather (1941) began his studies of what he called polygenic heredity. This was in spite of papers by the biometrician, Yule, in 1902 and 1906, showing that there was no irreconcilable difference between Galton's purely statistical law, essentially a multiple regression equation, and a physiological law such as Mendel's.

The situation was wholly different in America. Pearson's biometry (1901) was brought to America by Davenport, Harris and Pearl, but merely as a statistical tool. Castle not only published the first experimental Mendelian results from this country (1903a,b) but also soon began experimenting with inbreeding (Castle *et al.*, 1906) and selection, experimental population genetics. deVries had maintained that selection of quantitative variability had no permanent effect. This was probably because his artificial selection happened to be opposed by such strong natural selection that any progress was soon reversed on cessation of the artificial selection. This became something of a dogma, however, among many early Mendelians. Castle, with an agricultural background, challenged this dogma in an extensive experiment with black and white hooded rats, of which I was his assistant from 1912 to 1915. He selected toward self white and toward self black and was approaching these goals when high mortality and low fecundity put an end to both strains. Figure 2 shows the results of some 20 generations of selection in each direction.

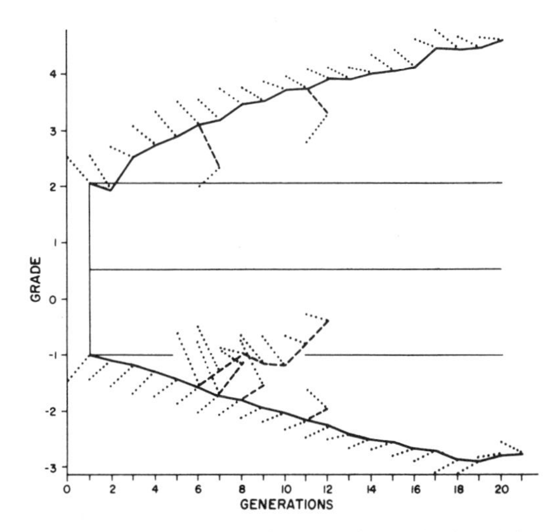

Figure 2. Courses of plus and minus selection (solid lines) and of reverse selection (broken lines) for amount of color in coats of hooded rats. The dotted lines show the regressions from the grades of the selected parents. From data of Castle and Phillips (1914) and Castle (1916).

Meanwhile Nilsson-Ehle (1909) in Sweden, G. H. Shull (1908) and E. M. East (1910) in this country were giving extensive experimental support to the interpretation of quantitative variability as due to multiple minor Mendelian differences. Castle had originally held that his successful selection experiments depended on variability of the major spotting factor of the rat, and in one case clearly demonstrated a mutation at this locus. Later, however, he made crucial tests that convinced him that the results depended largely on multiple independent modifiers (1919).

Darwin's theory that evolution is due primarily to mass selection of quantitative variability was put in mathematical form by Fisher in 1930 in his book, "The Genetical Theory of Natural Selection." According to his "fundamental theorem of natural selection": "the rate of increase in fitness of an organism at any time is equal to its genetic variance in fitness at that time." Evolution under given conditions according to this theorem, slows down to the exceedingly low rate supported by mutation as the genetic variance becomes exhausted. Long continued evolution at a higher rate depends on continued renewal of the genetic variance by changing conditions, which favor different alleles. Since he defined "genetic variance" as the additive component, dominance and gene interaction, the nonadditive components, merely reduce the rate. This theorem depended on Fisher's assumption that species are essentially homogeneous except in so far as different local environments may bring about selective differentiation. Fisher thus held that accidents of sampling, so called sampling drift, play no significant role apart from occasionally raising the frequency of a more favorable mutation to a safe level.

The shifting balance theory of evolution that I arrived at while in the Animal Husbandry Division of the U.S. Bureau of Animal Industry in Washington was very different from both Haldane's and Fisher's theories of the same period. It was based on my own experience in four very diverse research projects.

The first was Prof. Castle's selection experiment with hooded rats to which I have already referred.

The second was my thesis project (1916), studies of the genetics of several continuous series of variations in Prof. Castle's guinea pig colony; the intensities of the eumelanic and phaeomelanic colors, and patterns of hair direction, ranging from smooth to one of multiple small rosettes. There were independent loci in each case with allelic differences of various magnitudes, sets of multiple alleles, and different background heredities in different strains that showed apparent blending heredity, due presumably to multiple minor factors.

In the course of this study, I became fascinated with the frequently unpredictable effects of combinations. I had originally gone into genetics from interest in how genes act. The interaction effects seemed to give a basis for deducing the chains of gene action. Figure 3 shows the combination effects of the major genes on the eumelanic and phaeomelanic colors. Figure 4 shows the complex patterns of interaction of these genes and the multifactorial background heredities that seemed indicated. Figure 5 shows similarly those deduced in the case of hair direction.

The third project was the study of numerous closely inbred strains of guinea pigs which I

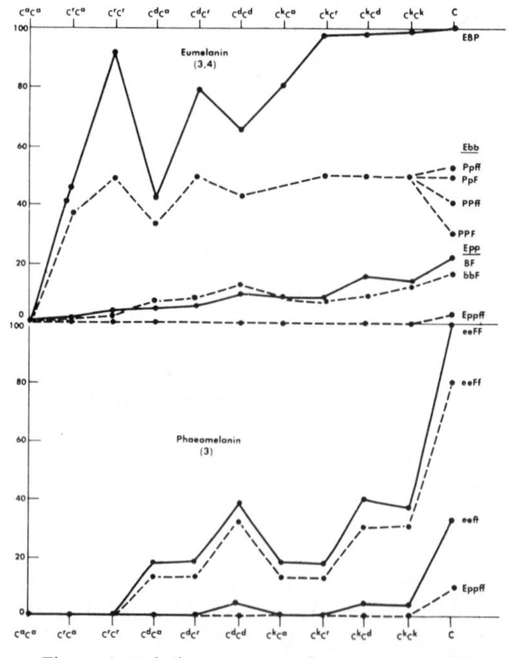

Figure 3. Relative amounts of pigment (eumelanin above, and phaeomelanin below) in coats of guinea pigs of various genotypes. (Wright, 1968, figure 5.8).

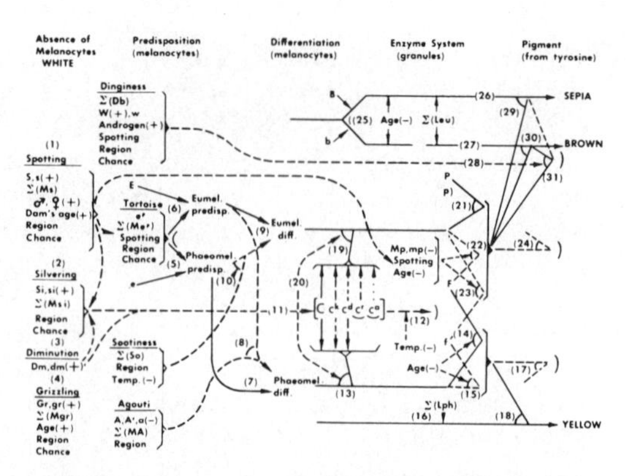

Figure 4. Factor interactions in the determination of coat color of the guinea pig. (Wright, 1968, figure 5.11).

took up on going to the Animal Husbandry Division in 1915. George Rommel, Chief of the Division, had started many strains from pairs in 1906 and had them maintained along lines of exclusive brother-sister mating. Twenty-three of these persisted long enough to yield significant data. Seventeen were still on hand when I was brought in to analyze the data and continue the experiment.

All strains exhibited the traditional inbreeding depression. There was not, however, a uniform decline in general vigor. Each strain had a unique combination of relative strength or weakness in size of litter and regularity in producing litters, mortality at birth and later, and in weight at birth and in later gains. Each strain had a particular combination of known color factors and its own background heredity with respect to the spotting pattern and intensity of color. There was also much differentiation in conformation. The very large animals of one strain (No. 13) (figure 6) had such short legs that they seemed to glide on the floor like

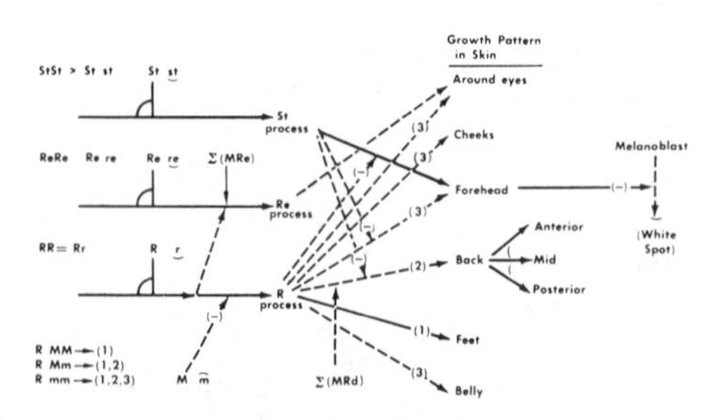

Figure 5. Diagram of factor interactions in the determination of hair direction and of a pleiotropic effect on a white forehead spot in the guinea pig. (Wright, 1968, figure 5.13).

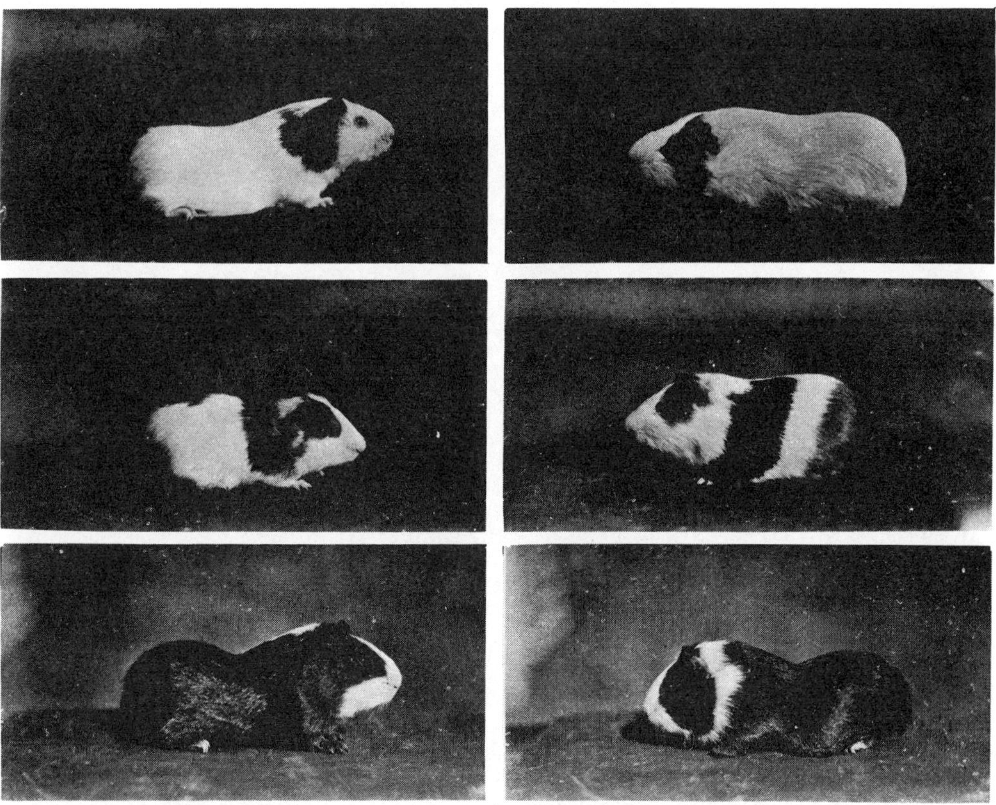

Figure 6. Illustration of varying color patterns and conformation in closely inbred strains of guinea pigs. Males (left) and females (right) of inbred strains 13, 2 and 39 (top to bottom) in generations 18, 12 and 13, respectively, of brother-sister mating. From Wright 1922c, plates II, I and V.

oversized planarians. The small animals of strain No. 2 had legs as long or longer than the preceding and ran well off the floor. Those of strain No. 13 had rounded noses and bent ears. Those of No. 2 had pointed noses and erect ears. Those of No. 2 had pointed noses and erect ears. Those of strains No. 39 had notably swayed backs. Strain 35 had protruding eyes; strain 13 sunken ones. The internal organs such as thyroid, adrenals and spleen differed strikingly in size and shape (Strandskov 1939, 1942). Some of the strains produced high frequencies of particular abnormalities such as otocephaly in No. 13, anophthalmia in No. 38. An atavistic little toe was common in No. 35, wholly absent in most others. There were notable differences temperament. The pigs of strain 13 could be picked up like sacks of meal while those of strains 2 and 35 would struggle and kick a hole in one's wrist unless picked up properly. Experiments by Dr. P. A. Lewis of the Phipps and later the Rockefeller Institute (Wright and Lewis, 1921), reveales striking differences in resistance to inoculations of tuberculosis in the order 35, 2, 13 and 32, 39 from high to low. Ones by Dr. Leo Loeb of Washington University, St. Louis, showed that each strain had its characteristic array of histocompatibility genes (Loeb and Wright, 1927).

My general project in the Animal Husbandry Division was clarification of the roles of inbreeding and selection in the breeding of livestock. This was the fourth of the projects referred to. I developed an inbreeding coefficient (the theoretical correlation between uniting gametes) that could be shown to measure the decrease in heterozygosis from that in the foundation stock. It could easily be derived, not only for any regular system of mating (Wright, 1921), but from the irregular ones of livestock pedigrees (Wright, 1922a) and by an approximation method, for whole breeds (Wright and McPhee, 1925; McPhee and Wright, 1925). Figure 7 shows the pedigree of the foundation Shorthorn bulls, Favourite and Comet. I calculated the coefficients for all of the 64 cows of Bates' famous Duchess line and their sires (Wright, 1923). Figure 8 shows the

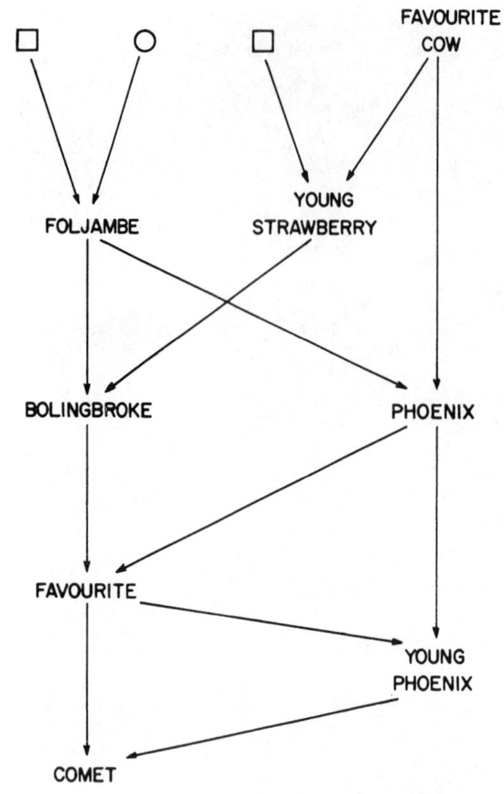

Figure 7. Pedigree of the Shorthorn bulls Favourite (252) and Comet (115). (Wright, 1977, figure 16.1).

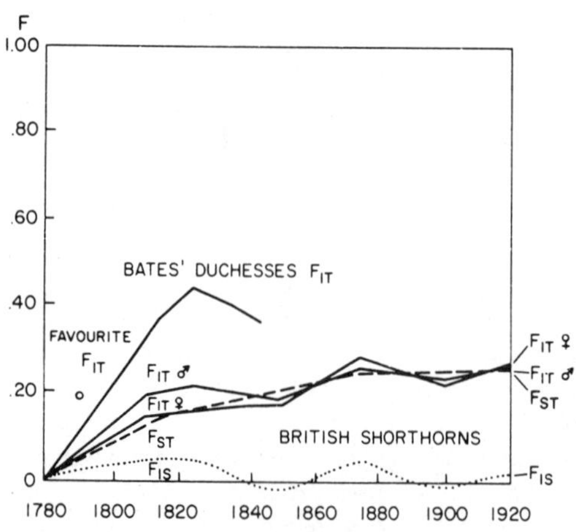

Figure 8. Inbreeding coefficients of British Shorthorn cattle from the foundation period 1780 to 1920. Inbreeding of bulls and cows relative to the foundation stock, F_{IT}, of the breed, and of the Bates' Duchesses, are in solid lines. Those for the hypothetical offspring of randomly mated animals, F_{ST}, are in a broken line. Those for individuals relative to the contemporary breed, F_{IS}, where $1 - F_{IT} = (1 - F_{IS})(1 - F_{ST})$ are in a dotted line. The inbreeding coefficient of the foundation bull, Favourite, is indicated by a circle. (Wright, 1977, figure 16.2).

results for the whole Shorthorn breed at intervals from about 1780, the foundation period, to 1920. This includes the coefficient F_{IT} (males and females separately) for inbreeding relative to the foundation stock, F_{ST}, derived by matching random two-line pedigrees of random sires and dams to find the cumulative inbreeding that would persist in spite of random breeding in the last generation, and F_{IS}, a derived coefficient, that measures current consanguine mating. The figure shows the strong inbreeding of Charles Colling's bull Favourite, born 1793, to which random animals in 1920, and doubtless 1977, were more closely related (.55) than offspring to sire under random mating. It also shows the extraordinarily high inbreeding of Bates' Duchesses, from the leading herd of the early 19th century. The relatively high values of current consanguine mating, F_{IS}, in the early years was followed by negative values in 1850. There was renewed excess inbreeding in 1875, but another dip at about 1900. These waves in F_{IS} undoubtedly represent alternate periods of development of strongly inbred herds and of prevailing crossing of animals as little related as possible within the breed.

I have described briefly the four rather unrelated lines of research that led me to a viewpoint on evolution very different from those propounded by Haldane and Fisher at about the same time, the 1920's. Let me indicate briefly, what it was that I derived from each of these lines.

From assisting Prof. Castle, I learned at firsthand the efficacy of mass selection in changing permanently a character subject merely to quantitative variability. Because of this and a distaste for miracles in science, I started with full acceptance of Darwin's contention that evolution depends mainly on quantitative variability rather than on favorable major mutations. Thus, I have assumed that species are typically heterallelic in tens of thousands of loci in which the leading alleles differ only slightly in effect, a situation that is maintained in a continually shifting state of near-equilibrium by the opposing pressures of recurrent mutation, diffusion and weak selection. Only a few loci at any time can show fairly rapid changes in allelic frequencies from strong selection.

My recognition of the efficacy of mass selection was, however, qualified by recognition that this process is likely to lead to deleterious side effects because of the interference of selection in any one respect with selection of the others.

From my studies of gene combinations, the second line, I recognized that an organism must never be looked upon as a mere mosaic of "unit characters," each determined by a single gene, but rather as a vast network of interaction systems. The indirectness of the relations of genes to characters insures that gene substitutions often have very different effects in different combinations and also multiple (pleiotropic) effects in any given combination. The latter consideration gives another reason for the tendency of mass selection to lead to deterioration.

The dependence of favorable interaction systems on multiple loci implies that if a selective value is attributed to each of the millions upon millions of possible sets of gene frequencies, the "surface" of selective values will have many peaks, each separated from the neighboring ones by saddles (figure 9). The elementary evolutionary process becomes passage from control by a relatively low selective peak to control by a higher one across a saddle, at first against the pressure of weak selection but under pressure of selection toward the new peak after the saddle has been passed. All of these considerations indicate that natural selection must somehow operate on combinations of interacting genes as wholes to be most effective. This raises a serious problem, however, since mass selection can operate only according to the net effects of substitutions under biparental reproduction (except for combinations of genes that are so closely linked that the combinations behave almost as if alleles).

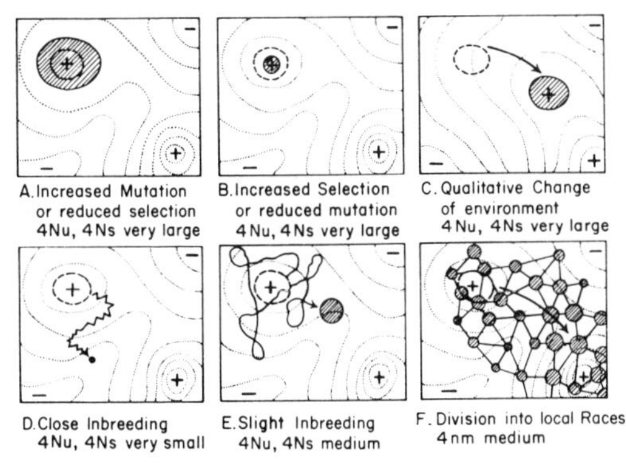

Figure 9. Hypothetical multidimensional field of gene frequencies (represented in two dimensions) with fitness contours. Field initially occupied by a population indicated by heavy broken contour except in case of multiple subdivisions in F. Field occupied later in A to E indicated by cross hatched area. Courses indicated in C, D, E, and F by arrows. Effective population number N (total), n (local). Coefficients u (mutation), s (selection), m (immigration). (Wright, 1977, figure 13.1, redrawn from Wright, 1932).

My studies of closely inbred lines of guinea pigs revealed the profound differentiations brought about in all respects by the cumulative accidents of sampling expected at all heterallelic loci under such inbreeding. Random fixation against the pressure of even rather strong selection under brother-sister mating accounts for the deterioration in random aspects. It shows, however, how a saddle leading to a higher selective peak may occasionally be crossed against the pressure of weak selection in small populations in which inbreeding is much less than under brother-sister mating. This suggested a solution to the difficulty of selection for interaction systems as wholes. There may often be subdivision of the species into small local populations that permit wide stochastic variability (but not fixation) at all nearly neutral loci and thus favor occasional local crossing of saddles leading to higher selective peaks. The firm establishment at this peak may then be brought about by mass selection. The spreading of this interaction system to neighboring localities may then be brought about by excess diffusion, followed again by mass selection after these have been pulled across the same saddle, and so on. The process may go on indefinitely as long as complete fixation does not occur. A major gene substitution and an array of modifiers that remove its deleterious side effects may also be established in this way.

It would have been very desirable to be able to point to studies of natural populations for evidence that such a process is actually occurring, or at least that the actual population structures are often favorable for its occurrence. Unfortunately there were few studies of differentiation within species except at the subspecies level at the time when I published the theory (1929, 1931). I had, however, been much impressed by Gulick's studies (1905) of what he called nonutilitarian differentiation in species of land snails in the numerous mountain valleys of Oahu in the Hawaiian Islands.

It was apparent, however, from my studies of the breeding history of Shorthorn cattle (the fourth line of research) that their improvement had actually occurred essentially by the shifting balance process rather than by mere mass selection. There were always many herds at any given time, but only a few were generally perceived as distinctly superior; those of Charles and Robert Collins near the end of the 18th century, those of Thomas Bates and the Booths in the first half of the 19th century, and that of Amos Cruickshank later in the century (figure 10). These herds successively made over the whole breed by being principal sources of sires.

My 1931 paper on evolution did not go into the factual side: the evidence for the multiple factor theory, for the importance of gene interaction and pleiotrophy or for the differentiation of inbred lines. Neither did I go into the exemplification on the process that I advocated in the history of breeds of livestock. I had discussed these elsewhere. The paper was almost wholly mathematical.

A

B

Figure 10. Prominent 19th Century Shorthorn sires. A: Cruickshank bull "Field Marshall" (47870). B: Bates' bull "Earl of Oxford, 3rd" (51186).

I attempted to introduce parameters for all of the factors into a single expression for the change of gene frequency, at least in token form, leading to an expression for the equilibrium frequency. Haldane had derived expressions for the change of gene frequency under various kinds of selection and integrated them to obtain curves describing the course toward fixation followed by the favorable allele under various conditions. Fisher's "fundamental theorem of natural selection" implies steady progress in exact accord with the steadily decreasing additive genetic variance as long as conditions remain unchanged. I did not derive any such expressions since under my view, evolution advanced by irregular, wholly unpredictable steps – the occasional occurrence off a peak-shift in some locality at some time, followed by relatively rapid spread throughout the species.

The mathematical emphasis in this paper was thus not on the course but on the momentary states of balance. The final summing up began as follows:

"Evolution as a process of cumulative change depends on a proper balance of the conditions which at each level of organization – gene, chromosome, cell, individual, local race – make for genetic homogeneity or genetic heterogeneity of the species."

After referring to the extensive field of variability always present in the state of balance at most loci, I went into the unlikelihood, under random mating, of evolution in very small populations or very large ones, or even in ones of any intermediate size. I stressed the great importance of subdivision into partially isolated local populations. I concluded in the final sentence that "conditions in nature are often such as to bring about the state of poise among opposing tendencies on which an indefinitely continuing evolutionary process depends."

I may have used the word "poise" in the final sentence partially because I had rather overworked the word "balance" (but not wholly). Balance may be maintained so firmly that further evolution is prevented. It itself becomes a product of evolution, an end result, as in most of the conspicuous polymorphisms. The word "poise" implies a lightly held state of balance at each moment, implying continual readiness to shift to a superior state of balance which is the essence of the shifting balance theory of evolution.

LITERATURE CITED

Bateson, W. 1909. Mendel's Principles of Heredity. Cambridge University Press, Cambridge.

Castle, W. E. 1903a. Mendel's Law of Heredity. Proc. Amer. Acad. Arts and Sci. 38.

Castle, W. E. 1903b. The laws of heredity of Galton and Mendel, and some laws governing race improvement by selection. Proc. Amer. Acad. Sci. 39:233.

Castle, W. E. 1916. Studies of inheritance in guinea-pigs and rats. Pub. No. 241. Carnegie Institution of Washington.

Castle, W. E. 1919. Piebald rats and selection: a correction. Amer. Naturalist 53:370.

Castle, W. E., F. W. Carpenter, A. H. Clark, S. O. Mast and W. M. Barrows. 1906. The effects of inbreeding, crossbreeding and selection upon the fertility and variability of Drosophila. Proc. Amer. Acad. Arts and Sci. 41:731.

Castle, W. E. and J. C. Phillips. 1914. Piebald rats and selection. Pub. No. 195. Carnegie Institution of Washington.

Darwin, C. 1859. The Origin of Species by Means of Natural Selection. John Murray, London.

East, E. M., 1910. A Mendelian interpretation of variation that is apparently continuous. Amer. Naturalist 44:65.

Fisher, R. A. 1930. The Genetical Theory of Natural Selection. Clarendon Press, Oxford.

Galton, F. 1889. Natural Inheritance. Macmillan and Co., London.

Gulick, J. T. 1905. Evolution, racial and habitudinal. Pub. No. 25. Carnegie Institution of Washington.

Haldane, J. B. S. 1924. A mathematical theory of natural and artificial selection. Part I. Trans. Cambridge Phil. Soc. 23:19.

Haldane, J. B. S. 1932. The Causes of Evolution. Longman, Green and Co., London.

Haldane, J. B. S. 1957. The Cost of Natural Selection. Genet. 55:511.

Loeb, Leo and S. Wright. 1927. Transplantation and individuality differentials in inbred strains of guinea pigs. Amer. J. Pathol. 31:251.

Mather, K., 1941. Variation and selection of polygenic characters. J. Genet. 41:159.

McPhee, H. C. and S. Wright. 1925. Mendelian analysis of the pure breeds of livestock, III The Shorthorns. J. Hered. 16:205.

Morgan, T. H. 1932. The Scientific Basis of Evolution. W. W. Norton and Co., Inc., New York.

Nilsson-Ehle, H. 1909. Kreuzungsuntersuchungen au Hafer und Weizen. Lunds Univ. Aerskr. N.F. 5:2:1–122.

Pearson, K. 1901. Mathematical contributions to the theory of evolution. VIII. On the inheritance of characters not capable of exact quantitative measurement. Part I. Introductory Part II. On the inheritance of coat-colour in horses. Part III. On the inheritance of eye-colour in man. Phil. Trans. Roy. Soc. London A 195:79.

Shull, G. H. 1908. The composition of a field of maize. Rep. Amer. Breeders Assoc. 4:290.

Strandskov, H. H. 1939. Inheritance of internal organ differences in guinea pigs. Genetics 24:722.

Strandskov, H. H. 1942. Skeletal variations in guinea pigs and their inheritance. J. Mammalogy 23:65.

Vries, Hugo de 1901–03. Die Mutationstheorie. 2 vol. Leipzig: Veit and Co.

Wright, S. 1916. An intensive study of the inheritance of color and other coat characters in guinea pigs with special reference to graded variation. Carnegie Institution of Washington: Pub. No. 241:59.

Wright, S. 1921. Systems of mating. Genetics 6:111.

Wright, S. 1922a. Coefficients of inbreeding and relationship. Amer. Naturalist 56:330.

Wright, S. 1922b. The effects of inbreeding and crossbreeding on guinea pigs. I. Decline in vigor. USDA Bull. 1090, p. 1. Washington, DC.

Wright, S. 1922c. Ibid. II. Differentiation among inbred families. USDA Bull. 1090, p. 37. Washington, DC.

Wright, S. 1922d. Ibid. III. Crosses between highly inbred families. USDA Bull. 1121. Washington, DC.

Wright, S. 1923. Mendelian analysis of the pure breeds of livestock. II. The Duchess family of Shorthorns as bred by Thomas Bates. J. Hered. 14:379.

Wright, S. 1929. Evolution in a Mendelian population. Anat. Rec. 44:87 (Abstr.).

Wright, S. 1931. Evolution in Mendelian populations. Genetics 16:97.

Wright, S. 1932. The roles of mutation, inbreeding, crossbreeding, and selection in evolution. Proc. VI. Intern. Cong. Genetics 1:356.

Wright, S. 1968. Evolution and the Genetics of Populations. Vol. I. Genetic and Biometric Foundations. Univ. of Chicago Press, Chicago, IL.

Wright, S. 1977. Evolution and the Genetics of Populations. Vol. III. Experimental Results and Evolutionary Deductions. Univ. of Chicago Press, Chicago, IL.

Wright, S. and P. A. Lewis. 1921. Factors in the resistance of guinea pigs to tuberculosis with especial regard to inbreeding and heredity. Amer. Naturalist 55:20.

Wright, S. and H. C. McPhee. 1925. An approximate method of calculating coefficients of inbreeding and relationship from livestock pedigrees. J. Agr. Res. 31:377.

Yule, G. U. 1902. Mendel's laws and their probable relation to intraracial heredity. New Phytol. 1:192, 222.

Yule, G. U. 1906. On the theory of inheritance of quantitative compound characters and the basis of Mendel's law: A preliminary note. Proc. III. Intern. Cong. Genetics, p. 140.

REFERENCES

Anonymous [Brewster, D.], 1838, Review of *Cours de Philosphie Positive* by August Comte, *Edinburgh Rev.* **67**:271–308 (July).

Bailey, L., 1906, *Plant-Breeding,* 4th ed., Macmillan, New York. (This edition contains an extensive bibliography. The first edition was published in 1896.)

Bajema, C., ed., 1976, *Eugenics: Then and Now,* Benchmark Papers in Genetics, vol. 5, Dowden, Hutchinson & Ross, Stroudsburg, Pa.

Barlow, N., ed., 1958, *The Autobiography of Charles Darwin, 1809–1882,* Harcourt, Brace and World, New York. (original omissions restored, appendix and notes added).

Bateson, W., 1894, *Material for the Study of Variation, Treated with Especial Regard to Discontinuity in the Origin of Species,* Macmillan, London.

Blyth, E., 1835, An Attempt to Classify the "Varieties" of Animals, with Observations on the Marked Seasonal Changes and Other Changes Which Naturally Take Place in Various British Species, and Which Do Not Constitute Varieties, *Mag. Nat. Hist.* **8**:40–53.

Bokonyi, S., 1974, *History of Domestic Mammals in Central and Eastern Europe, Akademiai Kiado, Budapest.*

Bowler, P. J., 1974, Darwin's Concept of Variation, *J. Hist. Med. Allied Sci.* **29**:196–212.

Buffon, G. comte de, 1769, *Histoire Naturelle Generale et Particuliere,* nouvelle ed., III, Paris.

Carlson, E. A., 1973, Eugenics Revisted: The Case for Germinal Choice, *Stadler Symp.* **5**:13–34.

Castle, W. E., 1905, The Mutation Theory of Organic Evolution from the Standpoint of Animal Breeding, *Science* **21**:521–525.

Castle, W. E., 1911, *Heredity in Relation to Evolution and Animal Breeding,* Appleton, New York.

Castle, W. E., 1919, Piebald Rats and Selection: A Correction, *Am. Nat.* **53**:370–376.

Castle, W. E., and H. MacCurdy, 1907, *Selection and Cross-Breeding in Relation to the Inheritance of Coat Pigments and Coat Patterns in Rats and Guinea Pigs,* Carnegie Institution of Washington Publ. No. 70, Washington, D.C.

Castle, W. E., and J. C. Phillips, 1914, *Piebald Rats and Selection,* Carnegie Institution of Washington Publ. No. 195, Washington, D.C.

Castle, W. E., and S. Wright, 1916, *Studies of Inheritance in Guinea-Pigs and Rats,* Carnegie Institution of Washington Publ. No. 241, Washington, D.C.

Chetverikov, S. S., 1926, On Certain Aspects of the Evolutionary Process from the Standpoint of Modern Genetics (in Russian), *Zh. Eksp. Biol.* **A2**:3–54. (An English translation appears in *Am. Philos. Soc. Proc.* **105**:167–195, 1961, and is reprinted as Paper 18 in Jameson, 1977*b*.)

Chetverikov, S. S., 1927, On the Genetic Constitution of Wild Populations (in Russian), *Int. Congr. Genetics, 1927, 5th*, Proc., pp. 1499–1500. (An English translation appears as Paper 19 in Jameson, 1977*b*.)

Clapperton, J., 1885, *Scientific Meliorism and the Evolution of Happiness*, Kegan Paul, Trench & Co., London, pp. 337–342. (These pages appear as Paper 6 in Bajema, 1976.)

Condorcet, M., 1795, *Esquisse d'un Tableau Historique des Progrès de l'Esprit Humain*, Masson, Paris. (*Sketch for a Historical Picture of the Progress of the Human Mind*, trans. J. Barraclough, 1955, Weidenfeld and Nicolson, London.)

Cook, R. C., 1932, Bacon Predicted Triumphs of Plant Breeding, *J. Hered.* **23**:162–165.

Cooper, J., 1808, Change of Seed Not Necessary to Prevent Degeneracy; Naturalization of Plants: Important Caution to Secure Good Quality of Plants, *Philadelphia Soc. Promot. Agric. Mem.* **1**:11–18 (appendix).

Culley, G., 1807, *Observations on Live Stock, Containing Hints for Choosing and Improving the Best Breeds of the Most Useful Kinds of Domestic Animals*, 4th ed., G. G. & J. Robinson, London. (The first edition was published in 1786.)

Curie-Cohen, M., L. Luttrell, and S. Shapiro, 1979, Current Practices of Artificial Insemination by Donor in the United States, *New England J. Med.* **300**:585–590.

Darwin, C. R., 1839, *Questions About the Breeding of Animals.* Reprinted in facsimile with an introduction by G. de Beer, Society for the Bibliography of Natural History, London, 1968.

Darwin, C. R., 1858, Extract from an Unpublished Work on Species, by. C. Darwin, Esq., Consisting of a Portion of a Chapter Entitled, "On the Variation of Organic Beings in a State of Nature; on the Natural Means of Selection; on the Comparison of Domestic Races and True Species," *Linnean Soc. London (Zool.) J. Proc.* **3**:46–50.

Darwin, C. R., 1859, *On the Origin of Species by Means of Natural Selection, or the Preservation of Favoured Races in the Struggle for Life*, 1st ed., J. Murray, London. (Facsimile edition by Harvard University Press, 1966.)

Darwin, C. R., 1868, *The Variation of Animals and Plants under Domestication*, 1st ed., 2 vols., J. Murray, London. (The second edition was published in 1875.)

Darwin, C. R., 1871, *The Descent of Man and Selection in Relation to Sex*, 1st ed., 2 vols., J. Murray, London. (The second edition was published in 1874.)

Darwin, E., 1800, *Phytologia: Or the Philosophy of Agriculture and Gardening*, J. Johnson, London.

Darwin, F., ed., 1899, *The Life and Letters of Charles Darwin*, 2 vols., Appleton, New York.

Darwin, F., and A. C. Seward, eds., 1903, *More Letters of Charles Darwin*, 2 vols., Appleton, New York.

de Beer, G., ed., 1958, *Evolution by Natural Selection*, Cambridge University Press, Cambridge. (Reprinted in 1971 by Johnson Reprint Corp.)

de Beer, G., ed., 1960–1967, *Darwin's Notebooks on Transmutation of Species*.

Part I. First Notebook (Notebook B) (July 1837–February 1838), *Br. Mus. (Nat. Hist.) Bull. Hist. Ser.* **2**(2):27–73 (1960).

Part II. Second Notebook (Notebook C) (February 1838–July 1838). *Br. Mus. (Nat. Hist.) Bull. Hist. Ser.* **2**(3):77–117 (1960).

Part III. Third Notebook (Notebook D) (July 15, 1838–October 2, 1838), *Br. Mus. (Nat. Hist.) Bull. Hist. Ser.* **2**(4):29–150 (1960).

Part IV. Fourth Notebook (Notebook E) (October 1838–July 10, 1939), *Br. Mus. (Nat. Hist.) Bull. Hist. Ser.* **2**(5):153–183 (1960).

Part V. Addenda and Corrigenda, *Br. Mus. (Nat. Hist.) Bull. Hist. Ser.* **2**(6):187–200 (1961)

Part VI. Pages Excised by Darwin, *Br. Mus. (Nat. Hist.) Bull. Hist. Ser.* **3**(5):129–176 (1967).

de Vries, H., 1901–1903, *Die Mutationstheorie. Versuche und Beobachtungen Über dier Enstehung der Arten im Pflanzenreiche*, 2 vols., Veit & Co., Leipzig. (English translation by J. Farmer and A. Darbishire, 1910, Open Court, Chicago; reprinted by Krause, New York, 1969.)

de Vries, H., 1904, *Species and Varieties: Their Origin by Mutation.* Open Court, Chicago.

de Vries, H., 1907, *Plant Breeding: Comments on the Experiments of Nilsson and Burbank*, Open Court, Chicago.

Dobzhansky, Th., 1937, *Genetics and the Origin of Species*, 1st ed., Columbia University Press, New York. (The second edition was published in 1951.)

Dudley, J. W., 1977, 76 Generations of Selection for Oil and Protein Percentage in *Maize, Int. Conf. Quant. Genet., 1976, Proc.*, Iowa State University Press, Ames, Iowa, pp. 459–473.

Dudley, J. W., R. Lambert, and D. Alexander, 1974, Seventy Generations of Selection for Oil and Protein Concentration in the Maize Kernel, in *Seventy Generations of Selection for Oil and Protein in Maize*, ed. J. Dudley, Crop Science Society of America, Madison, Wis.

Dunn, I. C., 1965, *A Short History of Genetics*, McGraw-Hill, New York.

East, E., 1910. A Mendelian Interpretation of Variation That Is Apparently Continuous, *Am. Nat.* **44**:65–82.

Fisher, R. A., 1918, The Correlation Between Relatives on the Supposition of Mendelian Inheritance, *R. Soc. Edinburgh Trans.* **52**:399–433.

Fisher, R. A., 1930, *The Genetical Theory of Natural Selection*, 1st ed., Clarendon Press, Oxford. (The second edition was published in 1959.)

Fraas, K. N., 1852, *Geschichte der Landwirtschaft, oder: Geschichtliche Übersicht der Fortschritte Landwirtschaftlicher Kenntniisse in den Letzen 100 Jahren*, J. O. Clave, Prague.

Frankel, O., 1974, Genetic Conservation: Our Evolutionary Responsibility, *Genetics* **78**:53–65.

Frankel, O., and J. Hawkes, eds., 1975, *Crop Genetic Resources for Today and Tomorrow*, Cambridge University Press, Cambridge.

Freeman, R. B., and P. J. Gautrey, 1969, Darwin's Questions About the Breeding of Animals with a Note on Queries About Expression, *Soc. Bibliog. Nat. Hist. J.* **5**:220–225.

Galton, F., 1865, Hereditary Talent and Character, Part I, *Macmillan's Mag.* **12**:157–166. (This article appears as Paper 1 in Bajema, 1976.)

Gärtner, C., 1849, *Versuche und Beobachtungen Über die Bastarderzeugung im Pflanzenreich*, Stuttgart.

Ghiselin, M. T., 1969, *The Triumph of the Darwinian Method*, University of California Press, Berkeley.

Glass, B., 1959, Maupertuis, Pioneer of Genetics and Evolution, in *Forerunners of Darwin: 1745–1859*, eds. B. Glass, O. Temkin, and W. Strauss, Johns Hopkins Press, Baltimore, Md., pp. 51–83.

Gruber, H., and W. Barrett, 1974,*Darwin on Man*, Dutton, New York.

Guyenot, E., 1941, *Les Sciences de la Vie aux XVIIe et XVIIIe Siecles: l'Idee d'Evolution*, Albin Michel, Paris.

Haldane, J. B. S., 1924–1932, A Mathematical Theory of Natural and Artificial Selection, Part I, *Cambridge Philos. Soc. Proc.* **23**:19–41; Part II, **23**:158–163; Part III, **23**:363–372; Part IV, **23**:607–615; Part V, **23**:838–844; Part VI, **26**:220–230; Part VII, **27**:131–136; Part VIII, **27**:137–142; Part IX, **28**:244–248. (Part VI appears as Paper 12 in Jameson, 1977*a* and Parts I, II, III, VII, VIII, and IX appear as Papers 15, 16, 17, 21, 22, and 23 in Jameson, 1977*b*.)

Haldane, J. B. S., 1932. *The Causes of Evolution*, Longmans, Green & Co., London. (Reprinted by Cornell Paperbacks, 1966.)

Hamilton, W., 1964, The Genetical Evolution of Social Behavior, *J. Theor. Biol.* **7**:1–52.

Harrison, F., ed., 1917, *Roman Farm Management: The Treatises of Cato and Varro . . .*, 2d ed., Macmillan, New York.

Hoffmann, H., 1869, *Untersuchungen sur Bestimmung des Werthes von Species und Varietät: ein Bietrage zur Kritik der Darwin'schen Hypothese*, L. F. Rikersche Buchhandlung, Giessen.

Huxley, T., 1860, Darwin on the Origin of Species, *Westminster Rev.*, April, pp. 295–310.

Huxley, T., 1863, *On Our Knowledge of the Causes of the Phenomena of Organic Nature. Being Six Lectures to Working Men at the Museum of Practical Geology*, Hardwicke, London. (Reprinted in Huxley, 1896.)

Huxley, T., 1896, *Darwiniana: Essays*, Appleton, New York.

Iizuki, R., Y. Swada, N. Nishima, and M. Ohi, 1968, The Physical and Mental Development of Children born Following Artificial Insemination, *Int. J. Fertil.* **13**:24–32.

Jameson, D., ed., 1977*a*, *Genetics of Speciation*, Benchmark Papers in Genetics, vol. 9, Dowden, Hutchinson & Ross, Stroudsburg, Pa.

Jameson, D., ed., 1977*b*, *Evolutionary Genetics*, Benchmark Papers in Genetics, vol. 8, Dowden, Hutchinson & Ross, Stroudsburg, Pa.

Jennings, H. S., 1908, Heredity, Variation, and Evolution in Protozoa: 2. Heredity and Variation of Size and Form in Paramecium, with Studies of Growth, Environmental Action, and Selection. *Am. Philos. Soc. Proc.* **47**:393–546.

Jennings, H. S., 1909, Heredity and Variation in the Simplest Organisms, *Am. Nat.* **43**:321–337.

Jennings, H. S., 1910, Experimental Evidence on the Effectiveness of Selection, *Am. Nat.* **44**:136–145.

Johannsen, W., 1903, *Ueber Erblichkeit in Populationen und in Reinen Linien*, Gustav Fischer, Jena.

Johannsen, W., 1907, Does Hybridization Increase Fluctuating Variability? *Report of the Third International Conference on Genetics*, Spottiswoode, London, pp. 98–113.

Kohn, D., 1980, Theories to Work by: Rejected Theories, Reproduction, and Darwin's Path to Natural Selection, *Stud. Hist. Biol.* **4**:67–170.

Lecoq, H., 1845, *De la fécondation naturelle at artificielle des végétaux, et de l'hybridation, considérée dans ses rapports avec l'horticulture, l'agriculture et la sylviculture, . . .,* 10th ed., Paris.

Le Couteur, J., 1836, *On the Varieties, Properties and Classification of Wheat*, Shearsmith, London.

Levin, D., ed., 1979, *Hybridization: An Evolutionary Perspective*, Benchmark Papers in Genetics, vol. 11, Dowden, Hutchinson & Ross, Stroudsburg, Pa.

Li, H., ed., 1977. *Stochastic Models in Population Genetics*, Benchmark Papers in Genetics, vol. 7, Dowden, Hutchinson & Ross, Stroudsburg, Pa.

Limoges, C., 1970, *La sélection naturelle. Étude sur la première constitution d'un concept (1837–1839)*, Presses Universitaires de France, Paris.

Lush, J. L., 1945, *Animal Breeding Plans*, 3d ed., Collegiate Press, Ames, Iowa. (The first edition was published in 1937.)

Lush, J. L., 1950, Genetics and Animal Breeding. in *Genetics in the Twentieth Century*, ed. L. C. Dunn, Macmillan, New York, pp. 493–525.

Lyell, C., 1837, *Principles of Geology, Being an Attempt to Explain the Former Changes of the Earth's Surface, by Reference to Causes Now in Operation*, vol. 2, 5th ed., J. Murray, London. (The first edition was published in 1830–1833.)

McKinney, H., 1969, Wallace's Earliest Observations on Evolution: 28 December 1845, *Isis* **60**:370–373.

McKinney, H., 1972, *Wallace and Natural Selection*, Yale University Press, New Haven, Conn.

Malthus, T. R., 1826, *An Essay on the Principle of Population, or, A View of Its Past and Present Effects on Human Happiness: With an Inquiry Into Our Prospects Regarding the Future Removal or Mitigation of the Evils Which It Occasions*, 2 vols., 6th ed., J. Murray, London.

Manier, E., 1978, *The Young Darwin and His Cultural Circle: A Study of Influences Which Helped Shape the Language and Logic of the First Drafts of the Theory of Natural Selection*, D. Reidel, Boston.

Marshall, W., 1790, *The Rural Oeconomy of the Midland Counties; Including the Management of Livestock in Leicestershire and Its Environs . . .,* 2 vols., G. Nicol, London.

Maupertuis, P.-L. M. de, 1745, *Venus Physique: Premiere Partie Contenant une Dissertation sur l'Origine des Hommes, et des Animaux; Second Partie Contenant une Dissertation sur l'Origine des Noirs.*

Maupertuis, P.-L. M. de, 1966, *The Earthly Venus*, trans. S. B. Boas, Johnson Reprint Corp., New York and London.

Maynard Smith, J., 1964, Group Selection and Kin Selection: A Rejoinder, *Nature* **201**:1145–1147.

Maynard Smith, J., 1965, Eugenics and Utopia, *Daedalus* **94**:487–505.

Mayr, E., 1942, *Systematics and the Origin of Species*, Columbia University Press, New York.

Mayr, E., 1959, Darwin and the Evolutionary Theory in Biology, in *Evolution and Anthropology: A Centennial Appraisal*, ed. B. Meggers, Anthropological Society of Washington, Washington, D.C., pp. 3–12.

Mayr, E., 1972, The Nature of the Darwinian Revolution, *Science* **176**:981–989.

Mayr, E., 1973, The Recent Historiography of Genetics, *J. Hist. Biol.* **6**:125–154.

Mayr, E., 1977, Darwin and Natural Selection, *Am. Sci.* **65**:321–327.

Morgan, T. H., For Darwin, *Pop. Sci. Mon.* **74**:367–380.

Muller, H. J., 1935, *Out of the Night: A Biologist's View of the Future*, Vanguard Press, New York.

Muller, H. J., 1959, The Guidance of Human Evolution, in *Evolution After Darwin*, vol. 2, ed. S. Tax, University of Chicago Press, Chicago, pp. 423–462. (This article appears as Paper 15 in Bajema, 1976.)

Muncy, R. L., 1973, *Sex and Marriage in Utopian Communities: Nineteenth Century America*, Indiana University Press, Bloomington, Ind.

Newman, L., 1912, *Plant Breeding in Scandinavia*, Canadian Seed Growers Association, Ottawa.

Nilsson-Ehle, H., 1909, Kreuzungsuntersuchungen an Hafer und Weizen, *Lund Univ. Aarsk. N.F., Afd.*, ser. 2, vol. 5, no. 2, 122pp.

Noyes, H. H., and G. W. Noyes, 1923, The Oneida Community Experiment in Stirpiculture, *in Eugenics, Genetics and the Family*, vol. 1, ed. C. Davenport et al., Williams and Wilkins, Baltimore, Md., pp. 374-386. (This article appears as Paper 5 in Bajema, 1976.)

Noyes, J. B., 1849, In *First Annual Report of the Oneida Association: Exhibiting Its History, Principles, and Transactions to January 1, 1849*, Leonard & Co., Oneida Reserve, New York.

Noyes, J. H., 1870, Scientific Propagation, in *The Modern Thinker*, vol. 1, ed. D. Goodman, American News Co., New York, pp. 97–120. (This article appears as Paper 4 in Bajema, 1976.)

Orel, V., 1977, Selection Practice and Theory of Heredity in Moravia Before Mendel, *Folia Mendeliana* **12**:179–200.

Orel, V., 1978, The Influence of T. A. Knight (1759–1838) on Early Plant Breeding in Moravia, *Folia Mendeliana* **13**:241–260.

Osborn, H., 1894, *From the Greeks to Darwin: An Outline of the Development of the Evolution Idea*, Macmillan, New York.

Osborn, H., 1912, Darwin's Theory of Evolution by the Selection of Minor Saltations, *Am. Nat.* **46**:76–82.

Pearl, R., 1911*a*, Breeding Poultry for Egg Production, *Maine Agric. Exp. Stn. Bull.* **192**:113–176.

Pearl, R., 1911*b*, Inheritance of Fecundity in Domestic Fowl, *Am. Nat.* **45**: 321–345.

Pearl, R., 1912, The Mode of Inheritance of Fecundity in the Domestic Fowl, *J. Exp. Zool.* **13**:153–268.

Pearl, R., 1917, The Selection Problem, *Am. Nat.* **51**:65–91.

Pearl, R., and F. Surface, 1909, Data on the Inheritance of Fecundity Obtained from the Records of Egg Production of the Daughters of "200-Egg" Hens, *Maine Agric. Exp. Stn. Bull.* **166**:49–84.

Pollak, E., O. Kempthorne, and T. Bailey, eds., 1977, *Proceedings of the*

International Conference on Quantitative Genetics, Iowa State University Press, Ames, Iowa.
Prichard, J. C., 1836–1847, *Researches into the Physical History of Mankind,* 5 vols., 3rd ed., John and Arthur Arch, London. (Previous editions were published in 1813 and 1826.)
Provine, W. B., 1971, *The Origins of Theoretical Population Genetics,* University of Chicago Press, Chicago.
Provine, W. B., 1978, The Role of Mathematical Population Geneticists in the Evolutionary Synthesis of the 1930s and 1940s, *Stud. Hist. Biol.* **2**: 167–192.
Rimpau, W., 1877, Züchtung neuer Getreidearten, *Landw. Jb.*
Rumpau, W., 1891, *Kreuzungsprodukte landwirtschaftlicher Kulturepflanzen,* Verlag P. Parey, Berlin, pp. 1–39.
Roberts, H. F., 1929, *Plant Hybridization Before Mendel,* Princeton University Press, Princeton, N.J.
Robertson, A., 1954, Artificial Insemination and Livestock Improvement, Adv. *Genetics* **6**:451–472.
Roll-Hansen, N., 1978, The Genotype Theory of Wilhelm Johannsen and Its Relation to Plant Breeding and the Study of Evolution, *Centaurus* **22**:210–235.
Roll-Hansen, N., 1980, The Controversy Between Biometricians and Mendelians: A Test Case for the Sociology of Scientific Knowledge, *Soc. Sci. Inf.* **19**:501–517.
Roper, A. G., 1913, *Ancient* Eugenics, Blackwell, Oxford. (Reprinted by Burgess, Minneapolis, 1975.)
Rudduck, H. B., ed., 1948, *Artificial Insemination of Farm Animals in the Soviet Union,* transl. J. S. Goode, Angus and Robertson, London.
Rümker, K. v., 1889, *Anleitung zur Getreidezüchtung auf Wissenschaftlicher und Praktischer Grundlage,* Verlag Paul Parey, Berlin.
Ruse, M., 1973, The Value of Analogical Models in Science, *Dialogue* **12:** 246–253.
Ruse, M., 1975, Charles Darwin and Artificial Selection, *J. Hist. Ideas* **36**: 339–350.
Ruse, M., 1979, *The Darwinian Revolution: Science Red Tooth and Claw,* University of Chicago Press, Chicago.
Schweber, S. S., 1977, The Origin of the *Origin* Revisited, *J. Hist. Biol.* **10**: 229–316.
Schweber, S. S., 1978, The Genesis of Natural Selection—1838: Some Further Insights, *BioScience* **28**:321–326.
Secord, J., 1981, Nature's Fancy: Charles Darwin and the Breeding of Pigeons, *Isis* 72:163–186.
Sharrock, R., 1672, *The History of the Propagation & Improvement of Vegetables by the Concurrence of Arts and Nature . . .,* 2d ed., R. Davis, Oxford.
Shirreff, P., 1873, *Improvement of the Cereals and an Essay on the Wheat-fly,* W. Blackwood, Edinburgh and London.
Simpson, G., 1944, *Tempo and Mode in Evolution,* Columbia University Press, New York.
Simpson, G., 1953, *The Major Features of Evolution,* Columbia University Press, New York.
Smith, A., 1976, *The Human Pedigree,* McGraw-Hill, New York.

Sprague, G. F., 1955, Problems in the Estimation and Utilization of Genetic Variability, *Cold Spring Harbor Symp. Quant. Biol.* **20**:87–92.
Stebbins, G., 1950, *Variation and Evolution in Plants,* Columbia University Press, New York.
Stubbe, H., 1972, *History of Genetics from Prehistoric Times to the Rediscovery of Mendel's Laws,* 2d ed., rev., M.I.T. Press, Cambridge, Mass.
"Student," 1933, Evolution by Selection: The Implications of Winter's Selection Experiment, *Eugenics Rev.* **24**:293–296.
Sturtevant, A. H., 1918, *An Analysis of the Effects of Selection,* Carnegie Institution of Washington Publ. No. 264, Washington, D.C., pp. 1–68.
Trow-Smith, R., 1957, *A History of British Livestock Husbandry to 1700,* Routledge & Kegan Paul, London.
Trow-Smith, R., 1959, *A History of British Livestock Husbandry: 1700–1900,* Routledge & Kegan Paul, London.
Verlot, B., 1865, *Sur la production et al fixation des variétés dans les plants d'ornement,* Paris.
Vilmorin, L., 1886, *Notices sur l'amélioriation des plantes par le semis, et considérations sur l'herédite dans les végétaux,* new ed., Paris.
Vorzimmer, P. J., 1969, Darwin's Questions About the Breeding of Animals, *J. Hist. Biol.* **2**:269–281.
Vorzimmer, P. J., 1970, *Charles Darwin: The Years of Controversy. The Origin of Species and Its Critics 1859–1882,* Temple University Press, Philadelphia.
Vorzimmer, P. J., 1977, The Darwin Reading Notebooks (1838–1860), *J. Hist. Biol.* **10**:107–153.
Walker, A., 1839, *Intermarriage: Or the Mode in Which, and the Causes Why, Beauty, Health and Intellect Result from Certain Unions and Deformity, Disease and Insanity from Others: Demonstrated by . . . and by an Account of Corresponding Effects in the Breeding of Animals,* London.
Wallace, A. R., 1858, On the Tendency Varieties to Depart Indefinitely from the Original Type, *Linn. Soc. London (Zool.) J. Proc.* **3**:53–62.
Wallace, A. R., 1866, July 2 Letter to Charles Darwin, in *More Letters of Charles Darwin,* vol. 1, ed. F. Darwin, Appleton, New York, 1903.
Wallace, A. R., 1903, My Relations with Darwin in Reference to the Theory of Natural Selection, *Black and White,* January 17, 1903, pp. 78–79.
Wallace, H., and W. Brown, 1956, *Corn and Its Early Fathers,* Michigan State University Press, East Lansing, Mich.
Weismann, A., 1909, The Selection Theory, in *Darwin and Modern Science,* ed. A. C. Seward, Cambridge University, Cambridge, pp. 18–65.
Wells, K. D., 1971, Sir William Lawrence (1783–1867): A Study of Pre-Darwinian Ideas on Heredity and Variation, *J. Hist. Biol.* **4**:319–361.
Wells, W., 1818, *Two Essays: One Upon Single Vision with Two Eyes; The Other on Dew . . . and an Account of the Female of the White Race of Mankind, Part of Whose Skin Resembles That of a Negro; with Some Observations on the Causes on the Differences in Colour and Form Between the White and Negro Races of Men,* Longman and Co., London.
Wilkie, J., 1956, The Idea of Evolution in Writings of Buffon, *Ann. Sci.* **12**: 48–62, 212–217, 244–266.

Wood, R. J., 1973, Robert Bakewell (1725–1795) Pioneer Animal Breeder His Influence on Charles Darwin, *Folia Mendeliana* **8**:231–242.

Wright, S., 1921, Systems of Mating, *Genetics* **6**:111–178.

Wright, S., 1931, Evolution in Mendelian Populations, *Genetics* **16**:97–159.

Wright, S., 1932, The Roles of Mutation, Inbreeding, Crossbreeding and Selection in Evolution, *Int. Congr. Genetics, 6th, Proc.* **1**:356–366.

Wright, S., 1968, *Evolution and the Genetics of Population,* vol. 1: *Genetic and Biometric Foundations,* University of Chicago Press, Chicago.

Wright, S., 1969, *Evolution and the Genetics of Populations,* vol. 2: *The Theory of Gene Frequencies,* University of Chicago Press, Chicago.

Wright, S., 1977, *Evolution and the Genetics of Populations,* vol. 3: *Experimental Results and Evolutionary Deductions,* University of Cicago Press, Chicago.

Wright, S., 1978, *Evolution and the Genetics of Populations,* vol. 4: *Variability Within and Among Natural Populations,* University of Chicago Press, Chicago.

Youatt, W., 1931, *The Horse: With a Treatise on Draught; and a Copius Index,* Baldwin and Cradock, London.

Youatt, W., 1834, *Cattle: Their Breeds, Management, and Diseases,* R. Baldwin, London.

Youatt, W., 1837, *Sheep: Their Breeds, Management, and Diseases,* Baldwin and Cradock, London.

Youatt, W., 1847, *The Pig: A Treatise on the Breeds, Management, Feeding and Medical Treatment of Swine,* Cradock, London.

Young, A., 1811, *On the Husbandry of Three Celebrated British Farmers, Messrs. Bakewell, Arbuthnot, and Ducket,* G. and W. Nicol. London.

Zirkle, C., 1935, *The Beginnings of Plant Hybridization,* University of Pennsylvania Press, Philadelphia.

AUTHOR CITATION INDEX

SUBJECT INDEX

About the Editor

CARL JAY BAJEMA is professor of biology in the College of Arts and Sciences at Grand Valley State Colleges, Allendale, Michigan. He teaches courses in human ecology, human genetics, human sexuality, evolutionary biology, and sociobiology in the Department of Biology, and the Darwinian revolution in the History of Science program. Professor Bajema received the Ph.D. in zoology from Michigan State University in 1963.

Professor Bajema has served as Senior Population Council Fellow in Demography and Population Genetics at the University of Chicago (1966–1967); as Research Associate in Population Studies at Harvard University (1967–1972); as Research Associate in Biology at the University of California, Santa Barbara (1974); and most recently as Visiting Professor of Anthropology at Harvard University (1974–1975).

Dr. Bajema's research interests have concentrated on the estimation of the direction and intensity of natural selection with respect to human behavior. Professor Bajema is editor of *Natural Selection in Human Populations* (Wiley, 1971) and co-author with Garrett Hardin of the third edition of the introductory college biology textbook *Biology: Its Principles and Implications* (Freeman, 1978). He is writing a monograph on the history of selection theory and is also editing additional volumes on selection theory for the *Benchmark Papers in Systematic and Evolutionary Biology* series being published by Hutchinson Ross Publishing Company.

Professor Bajema enjoys hiking, bicycling and cross-country skiing. He resides in Grand Rapids, Michigan.